Dimensions Math®
Textbook KA

Authors and Reviewers

Tricia Salerno

Pearly Yuen

Jenny Kempe

Cassandra Turner

Elizabeth Curran

Allison Coates

Consultant

Dr. Richard Askey

Singapore Math Inc.

Published by Singapore Math Inc.

19535 SW 129th Avenue
Tualatin, OR 97062
www.singaporemath.com

Dimensions Math® Textbook Kindergarten A
ISBN 978-1-947226-02-9

First published 2017
Reprinted 2018, 2019, 2020 (twice)

Printed in China

Acknowledgments

Editing by the Singapore Math Inc. team.
Design and illustration by Cameron Wray.

Preface

The Dimensions Math® Pre-Kindergarten to Grade 5 series is based on the pedagogy and methodology of math education in Singapore. The curriculum develops concepts in increasing levels of abstraction, emphasizing the three pedagogical stages: Concrete, Pictorial, and Abstract. Each topic is introduced, then thoughtfully developed through the use of exploration, play, and opportunities for mastery of skills.

Features and Lesson Components

Students work through the lessons with the help of five friends: Emma, Alex, Sofia, Dion, and Mei. The characters introduce themselves in Pre-K and continue to appear throughout the series. They give instructions, hints, and ideas.

Chapter Opener

Each chapter begins with an engaging scenario that stimulates student curiosity in new concepts. This scenario also provides teachers an opportunity to review skills.

Lesson

Engaging pictures draw the students into the concept of each lesson.

Exercise

A pencil icon at the end of the lesson links to additional practice problems in the workbook.

Review

A review of chapter material provides ongoing practice of concepts and skills.

Note: There are additional lesson components in the teacher's guide: Explore, Learn, Play, and Extend.

Contents

Chapter		Lesson	Page
Chapter 1 Match, Sort, and Classify		Chapter Opener	1
	1	Left and Right	2
	2	Same and Similar	4
	3	Look for One That is Different	6
	4	How Does it Feel?	8
	5	Match the Things That Go Together	10
	6	Sort	12
	7	Practice	13
Chapter 2 Numbers to 5		Chapter Opener	15
	1	Count to 5	16
	2	Count Things Up to 5	19
	3	Recognize the Numbers 1 to 3	22
	4	Recognize the Numbers 4 and 5	25
	5	Count and Match	27
	6	Write the Numbers 1 and 2	29
	7	Write the Number 3	32
	8	Write the Number 4	34
	9	Trace and Write 1 to 5	36
	10	Zero	38
	11	Picture Graphs	42
	12	Practice	45

Chapter		Lesson	Page
Chapter 3 **Numbers to 10**		Chapter Opener	49
	1	Count 1 to 10	50
	2	Count Up to 7 Things	51
	3	Count Up to 9 Things	53
	4	Count Up to 10 Things — Part 1	56
	5	Count Up to 10 Things — Part 2	59
	6	Recognize the Numbers 6 to 10	60
	7	Write the Numbers 6 and 7	62
	8	Write the Numbers 8, 9, and 10	64
	9	Write the Numbers 6 to 10	67
	10	Count and Write the Numbers 1 to 10	69
	11	Ordinal Positions	72
	12	One More Than	75
	13	Practice	78

Chapter		Lesson	Page
Chapter 4 **Shapes and Solids**		Chapter Opener	81
	1	Curved or Flat	82
	2	Solid Shapes	83
	3	Closed Shapes	85
	4	Rectangles	86
	5	Squares	88
	6	Circles and Triangles	89
	7	Where is It?	92
	8	Hexagons	94
	9	Sizes and Shapes	96
	10	Combine Shapes	98
	11	Graphs	100
	12	Practice	102
Chapter 5 **Compare Height, Length, Weight, and Capacity**		Chapter Opener	105
	1	Comparing Height	106
	2	Comparing Length	109
	3	Height and Length — Part 1	112
	4	Height and Length — Part 2	114
	5	Weight — Part 1	116
	6	Weight — Part 2	118
	7	Weight — Part 3	119
	8	Capacity — Part 1	121
	9	Capacity — Part 2	123
	10	Practice	125

Chapter		Lesson	Page
Chapter 6 **Comparing Numbers Within 10**		Chapter Opener	127
	1	Same and More	128
	2	More and Fewer	130
	3	More and Less	133
	4	Practice — Part 1	137
	5	Practice — Part 2	141

Chapter 1

Match, Sort, and Classify

Lesson 1
Left and Right

1

Objective: Distinguish left and right.

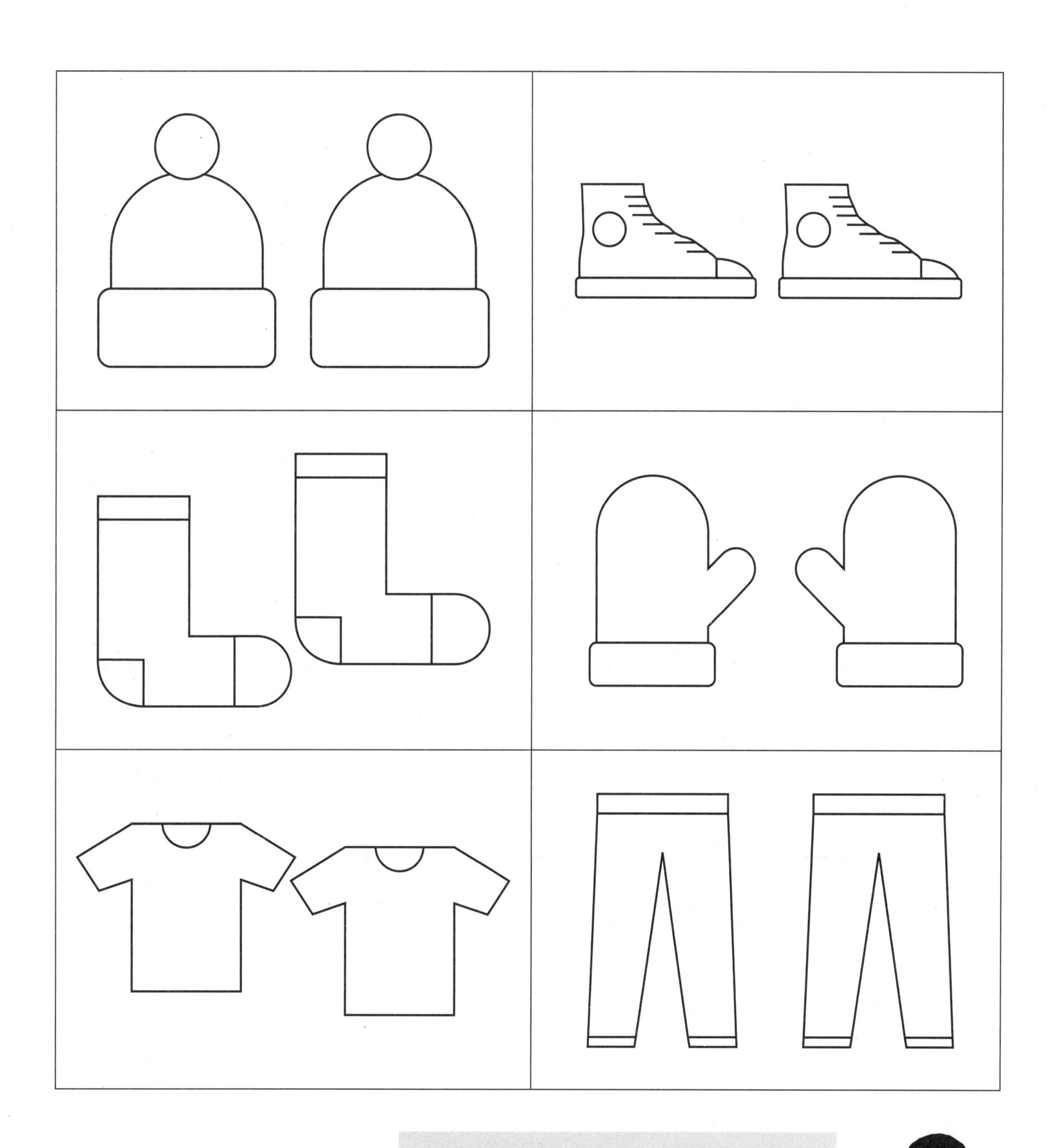

Color the left one of each object red and the right one green.

Objective: Distinguish left and right.

Exercise 1 • page 1

Lesson 2
Same and Similar

2

Objective: Find pairs of objects that are the same and that are similar.

Match the things that are the same or similar.

Objective: Match objects that are the same or similar.

Exercise 2 • page 3

Lesson 3
Look for One That is Different

3

Look and talk.
Which one is different?

Objective: Identify the objects that are different.

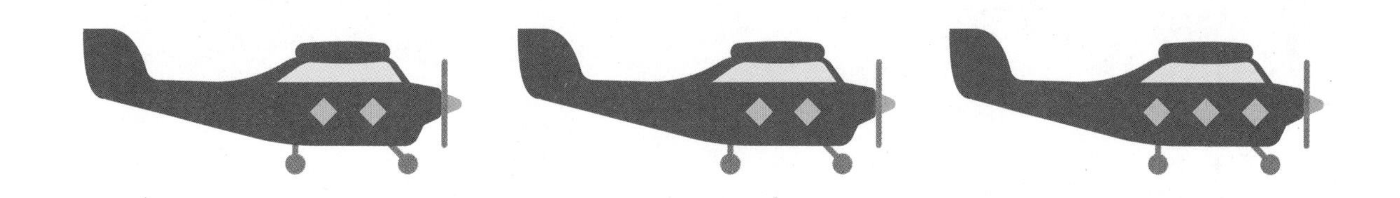

Which one is different?

Objective: Identify the objects that are different.

Exercise 3 • page 5

Lesson 4
How Does it Feel?

4

Objective: Describe an object as feeling smooth or rough.

Draw lines from rough things to the peanut, and from smooth things to the ribbon.

Objective: Determine if an object feels smooth or rough.

Exercise 4 • page 7

Lesson 5
Match the Things That Go Together

5

Objective: Identify objects that go together.

Objective: Match objects that go together and explain why.

Exercise 5 • page 9

Lesson 6
Sort

Cross out the button that doesn't belong and be ready to tell why.

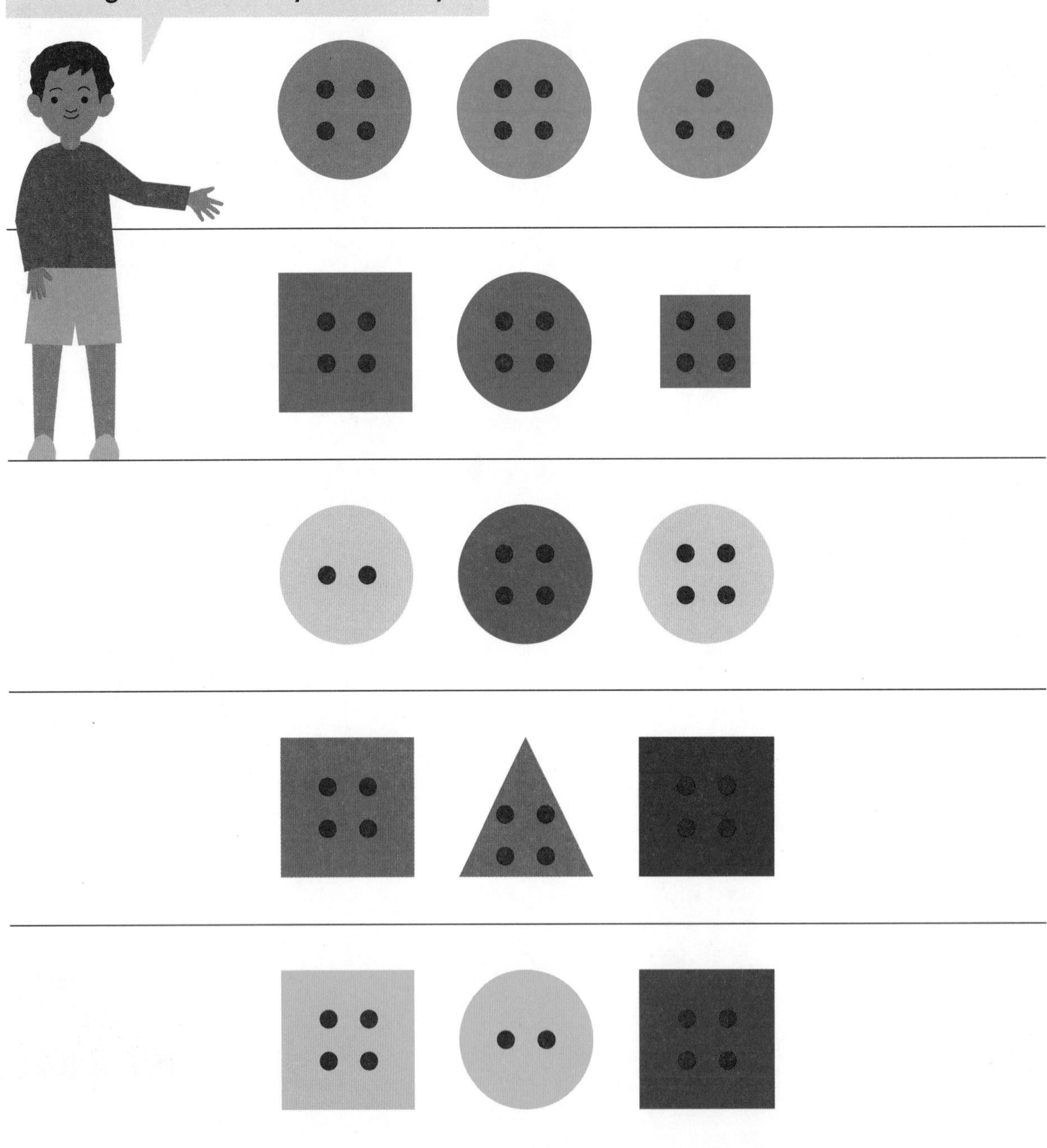

Objective: Sort objects in more than one way.

Exercise 6 • page 11

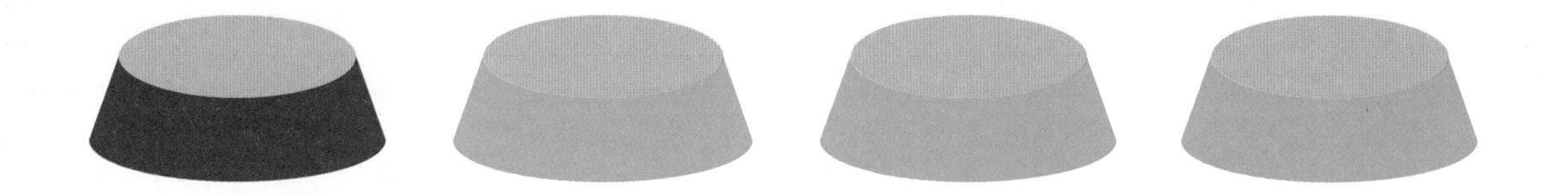

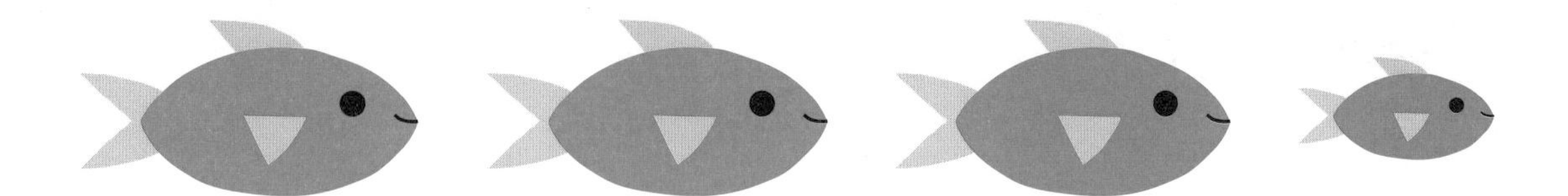

Cross out the one that is different.

Objective: Practice.

Draw a line to match things that go together.

Objective: Practice.

Exercise 7 • page 13

Chapter 2

Numbers to 5

Lesson 1
Count to 5

1

Objective: Count 1 to 5.

Objective: Count to 5.

Objective: Count to 5.

Exercise 1 • page 15

Lesson 2
Count Things Up to 5

2

Objective: Create sets and count the number of objects in the set.

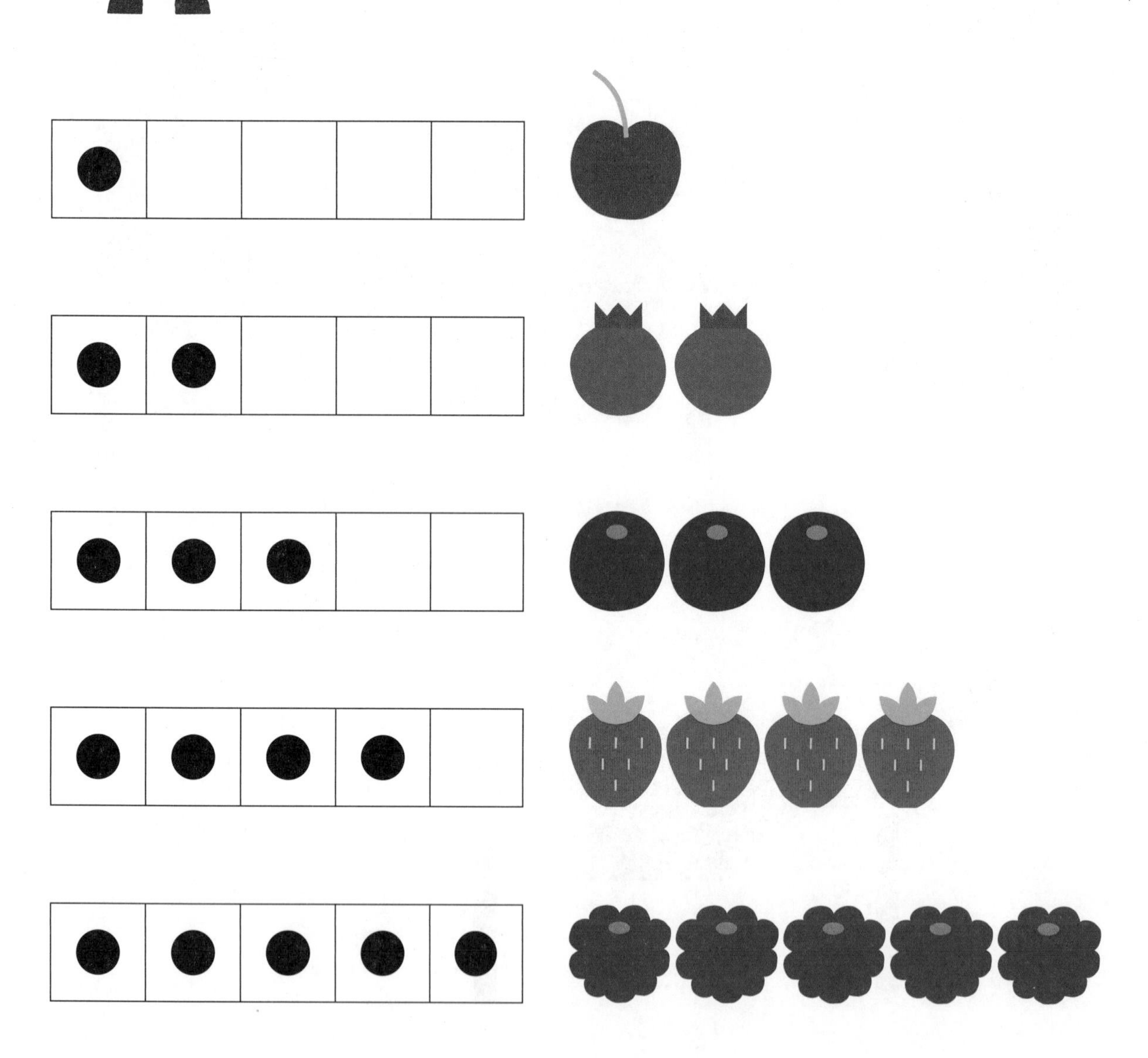

Objective: Recognize the five-frame cards that match a set of 1 to 5 objects.

Objective: Count and put the same number of counters on a five-frame card for each set of objects.

Exercise 2 • page 17

Lesson 3
Recognize the Numbers 1 to 3

3

Objective: Recognize the numerals 1, 2, and 3.

1
2
3

Draw circles to match each number.

Objective: Recognize the numerals 1, 2, and 3.

3

1

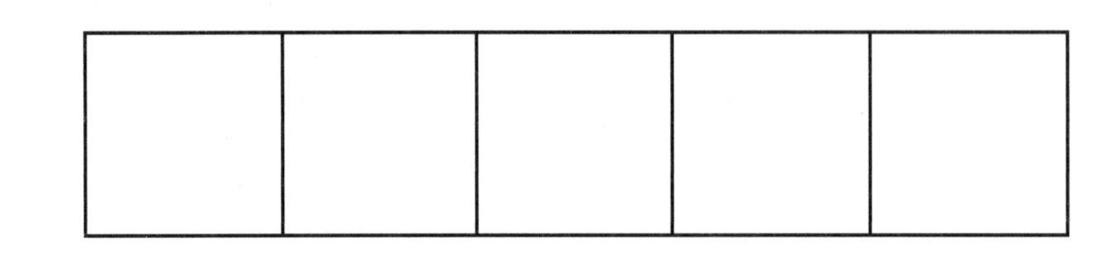

2

Color in the boxes to show the number.

Objective: Represent the numerals 1, 2, and 3 on a five-frame card.

Exercise 3 • page 19

Lesson 4
Recognize the Numbers 4 and 5

4

Objective: Count and match a set of 4 or 5 objects to the numerals.

Objective: Recognize the numerals 4 and 5.

Exercise 4 • page 21

Lesson 5
Count and Match

5

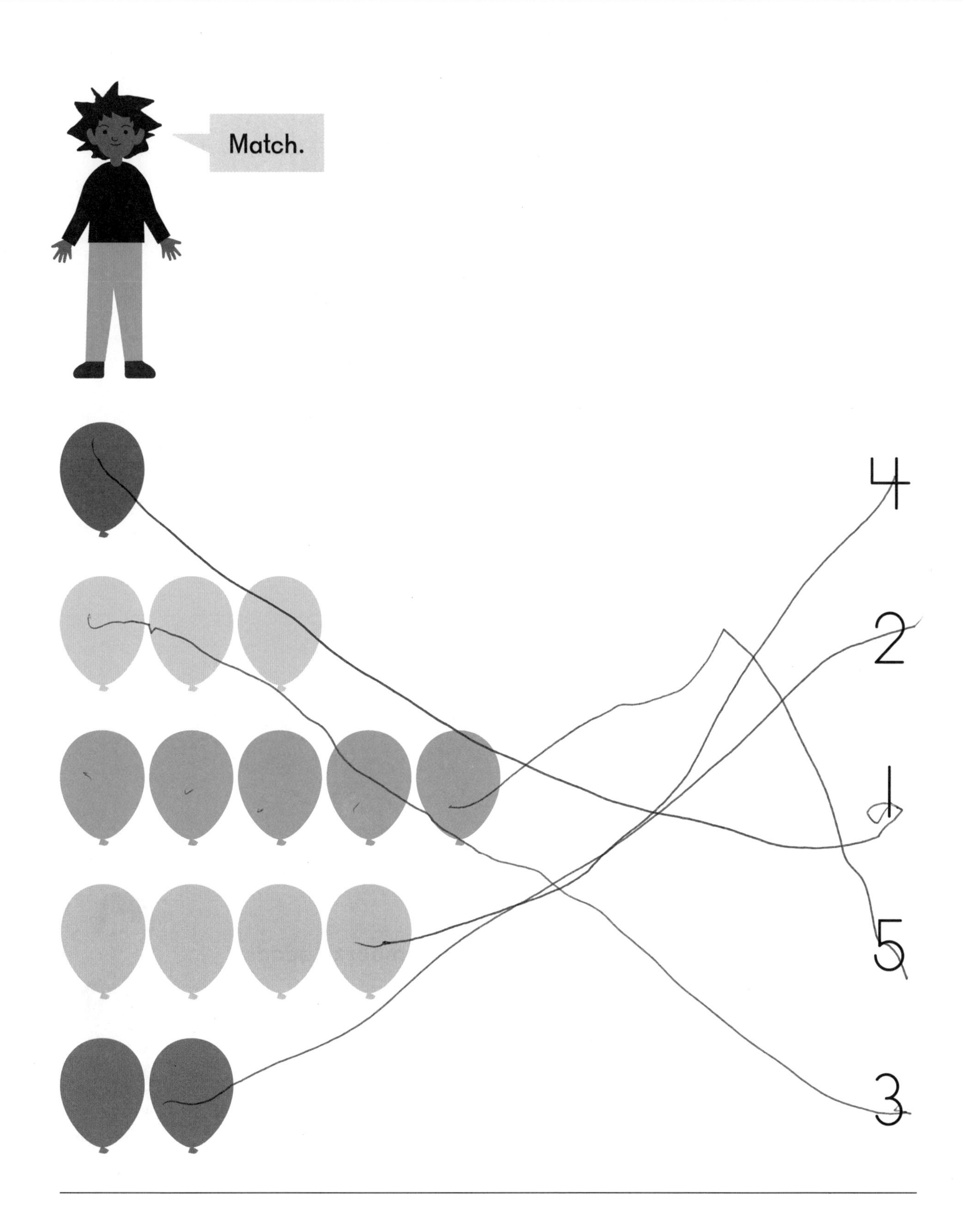

Objective: Recognize the numerals 1 to 5.

4

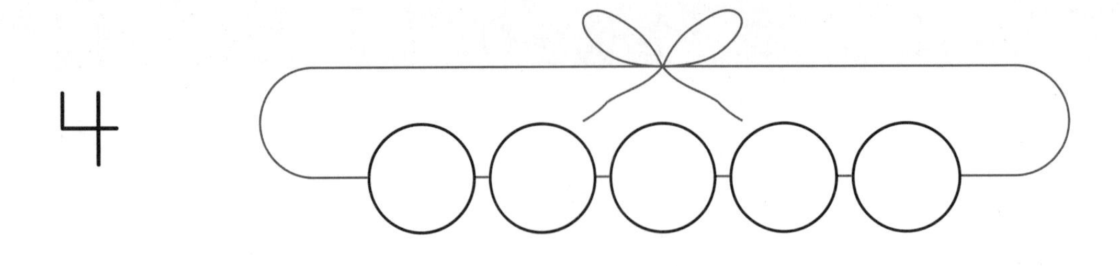

2

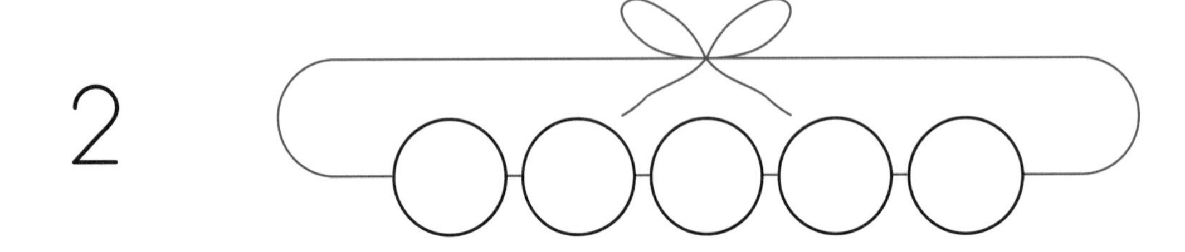

1

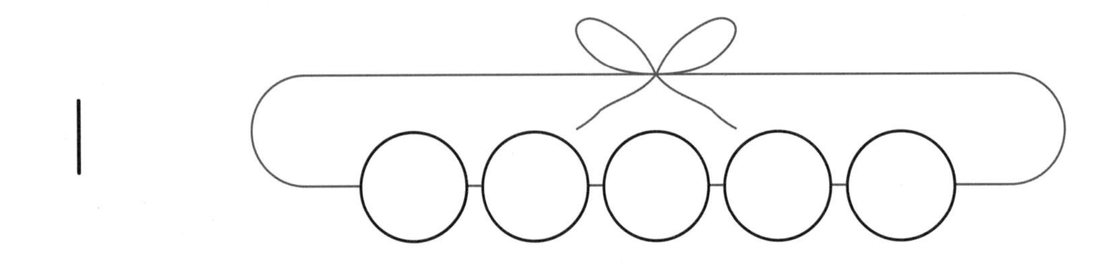

5

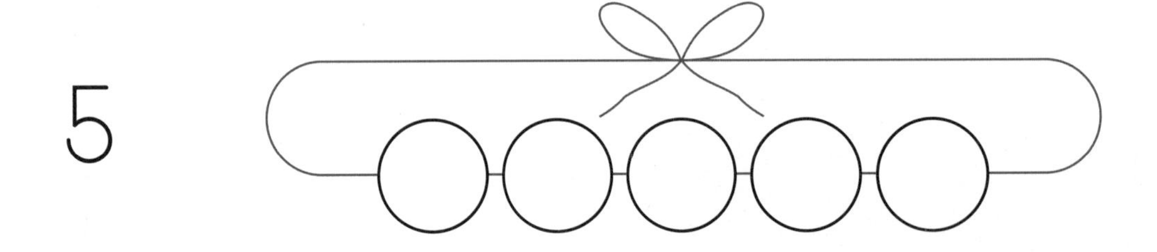

Color the correct number of beads.

Objective: Recognize the numerals 1 to 5.

Exercise 5 • page 23

Objective: Trace straight lines.

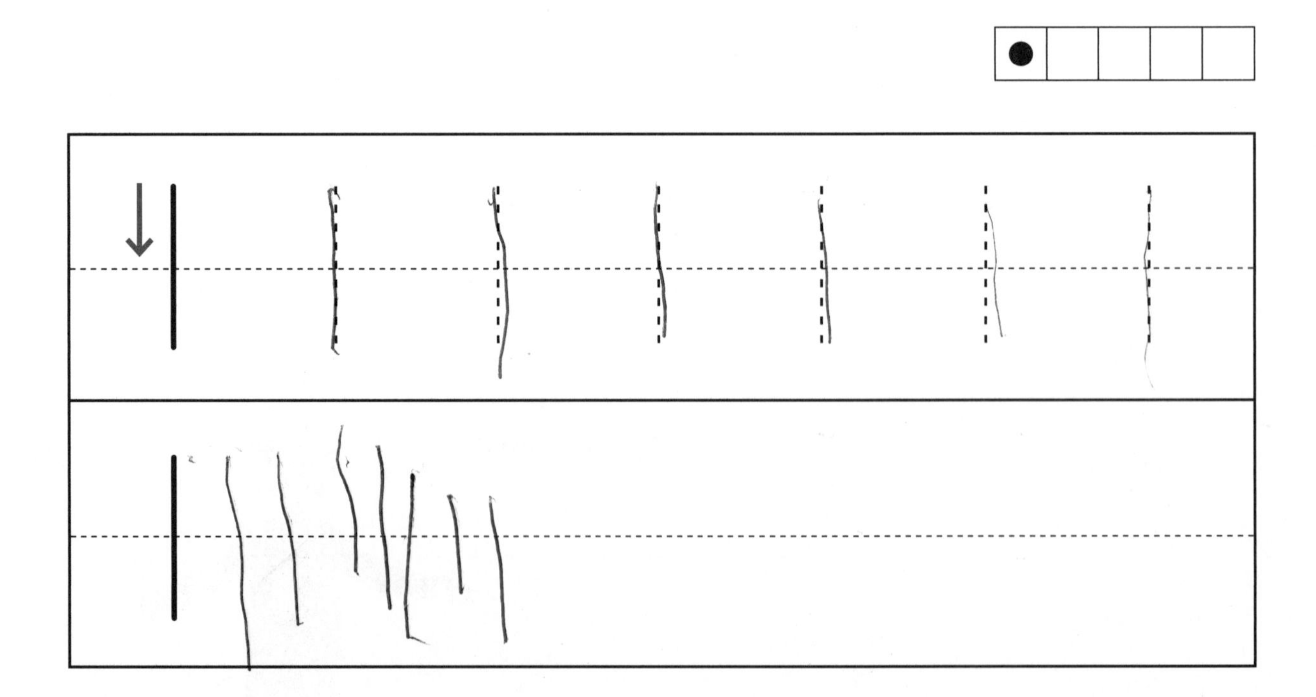

Objective: Trace straight lines and write the numeral 1.

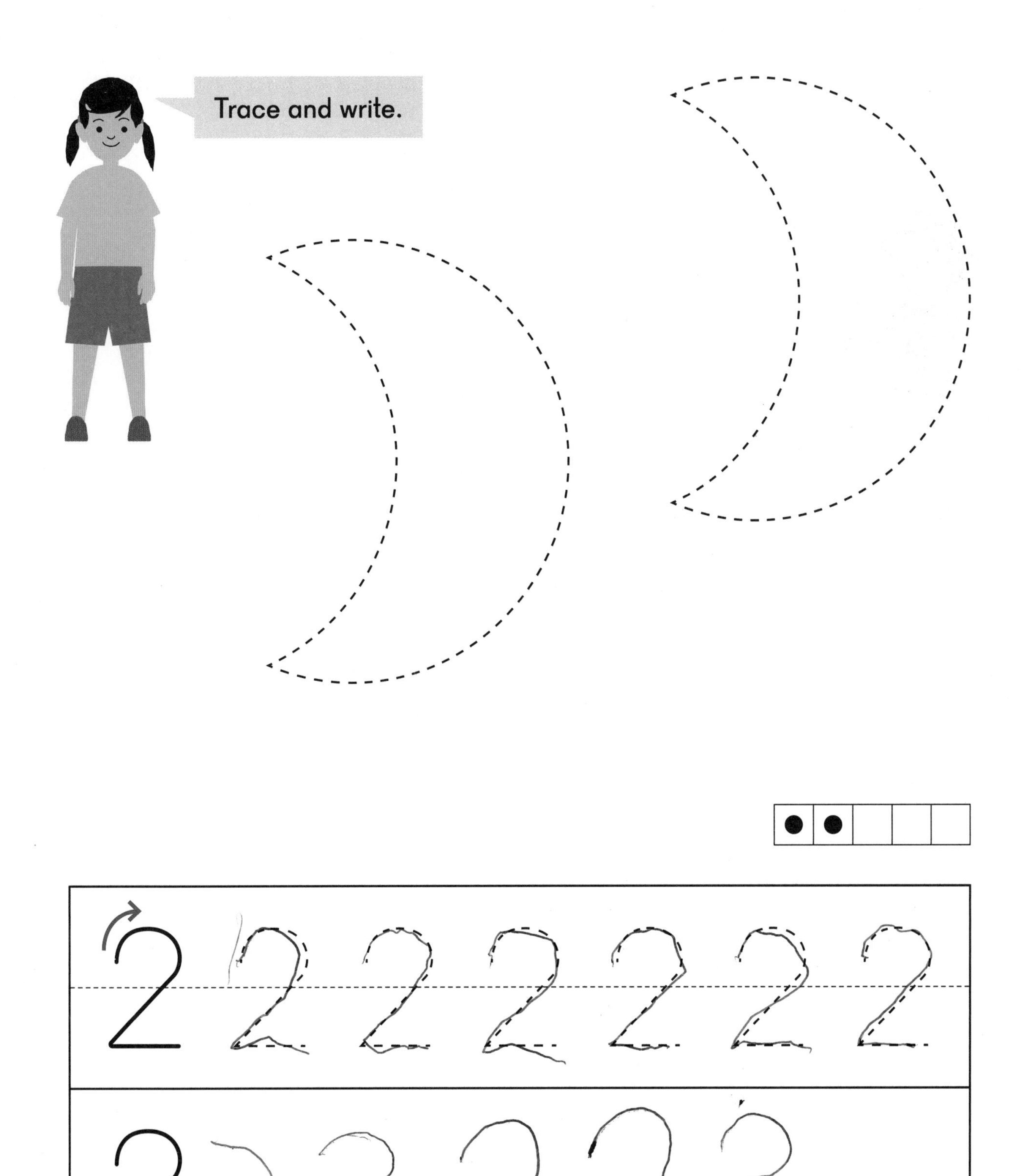

Objective: Trace curved lines and write the numeral 2.

Exercise 6 • page 27

Write the Number 3

Objective: Trace curved lines and write the numeral 3.

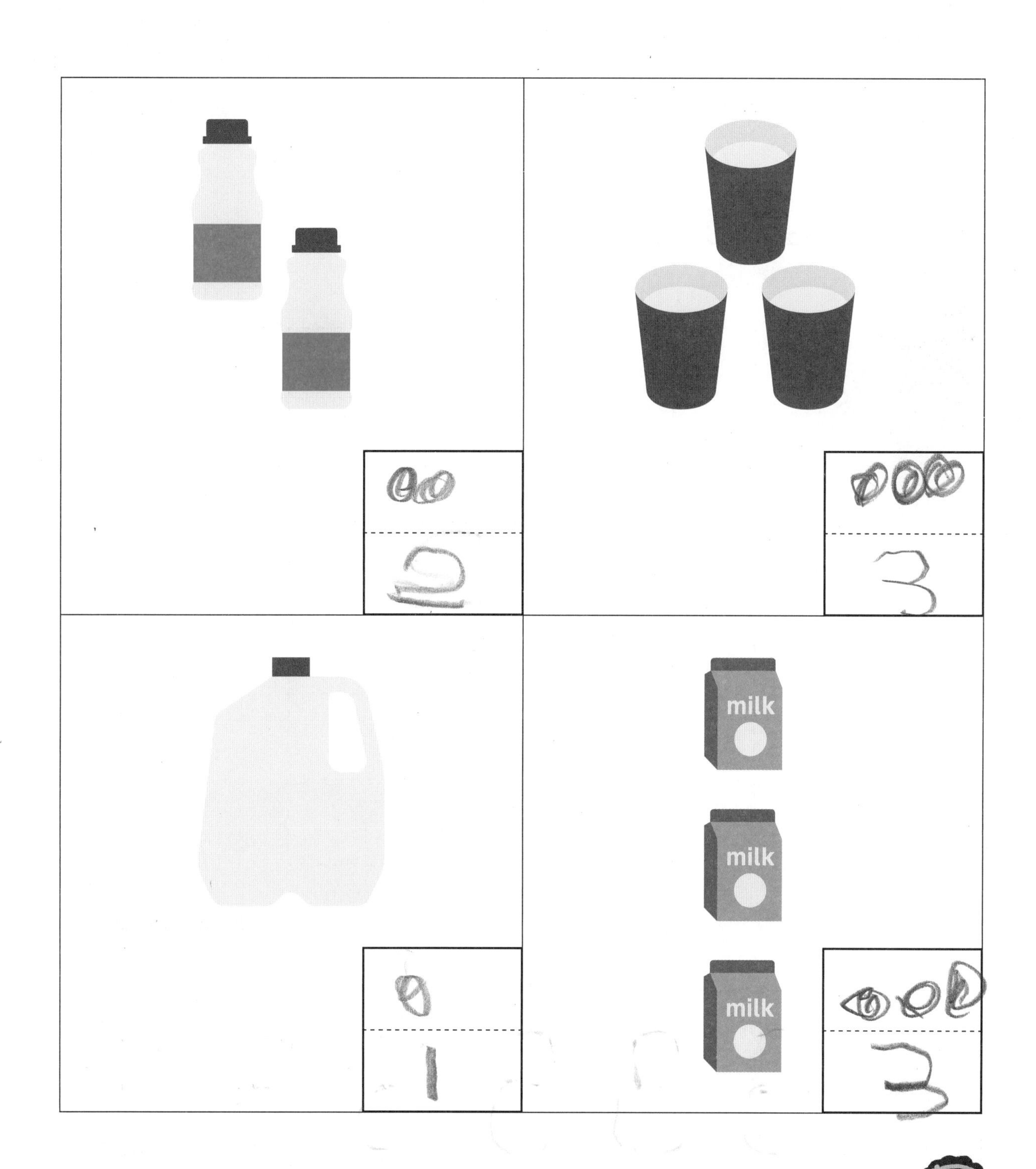

Count and write the numbers.

Objective: Write the numerals 1, 2, and 3 to match a set containing that number of objects.

Exercise 7 • page 29

Lesson 8
Write the Number 4

8

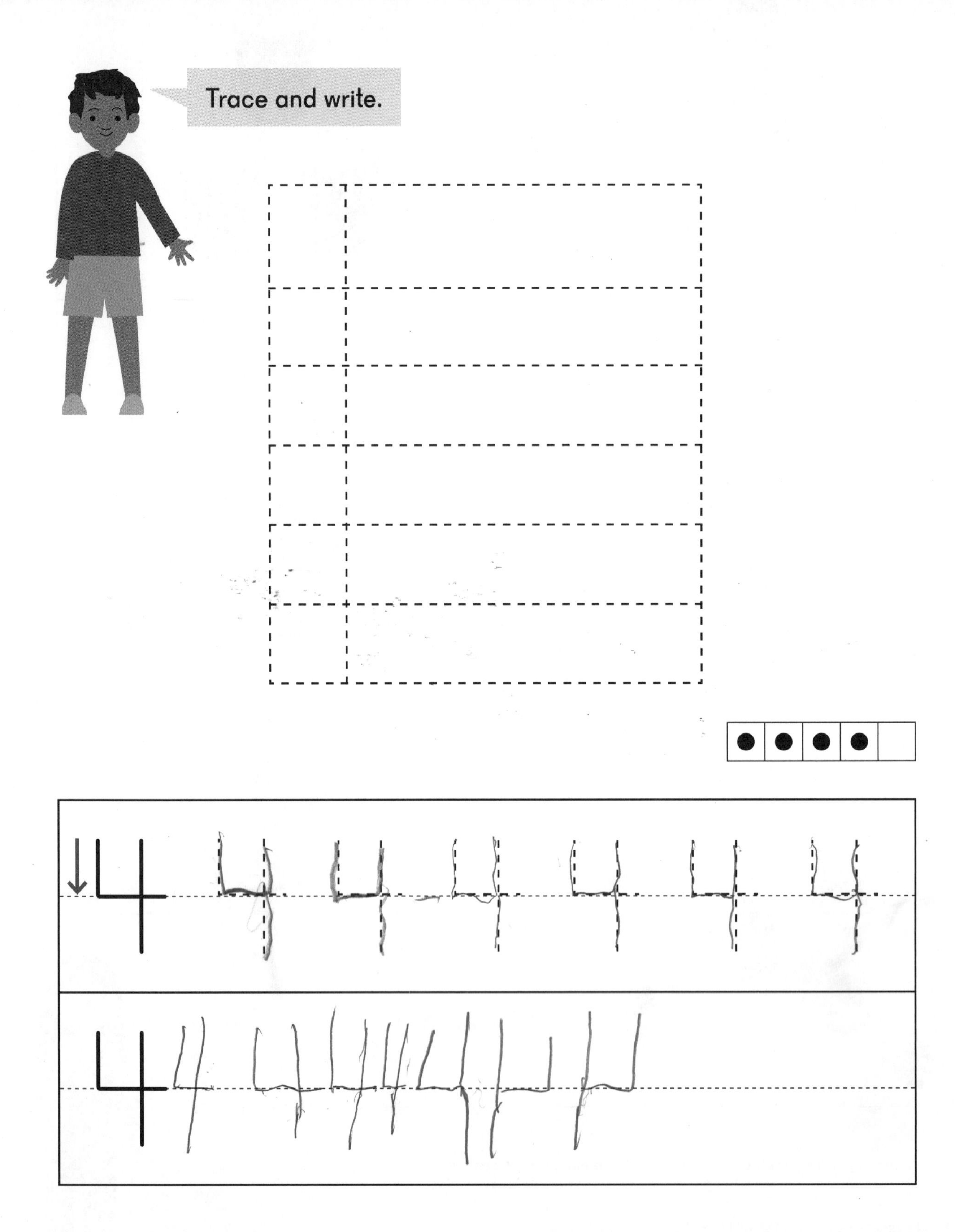

Objective: Write the numeral 4.

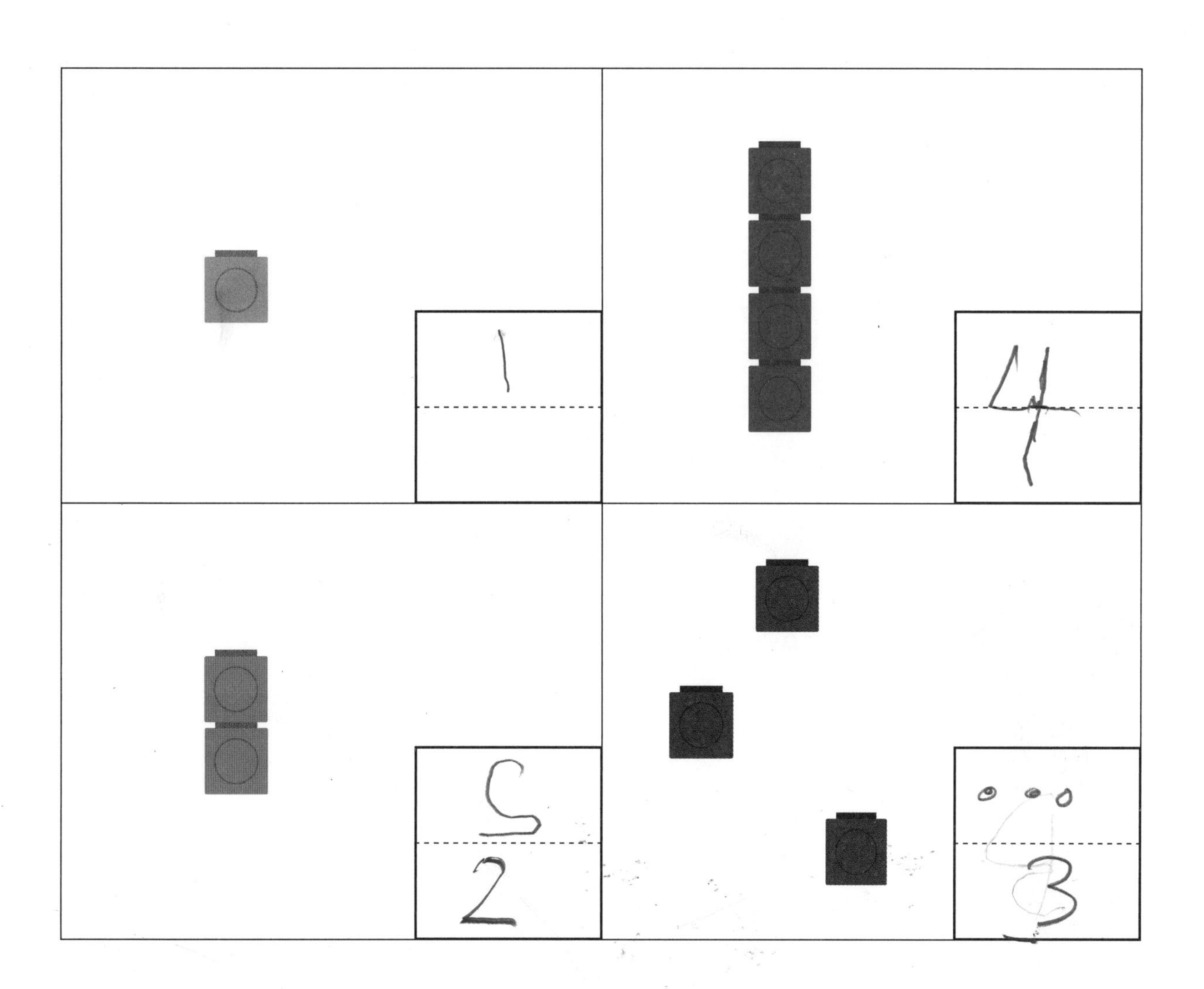

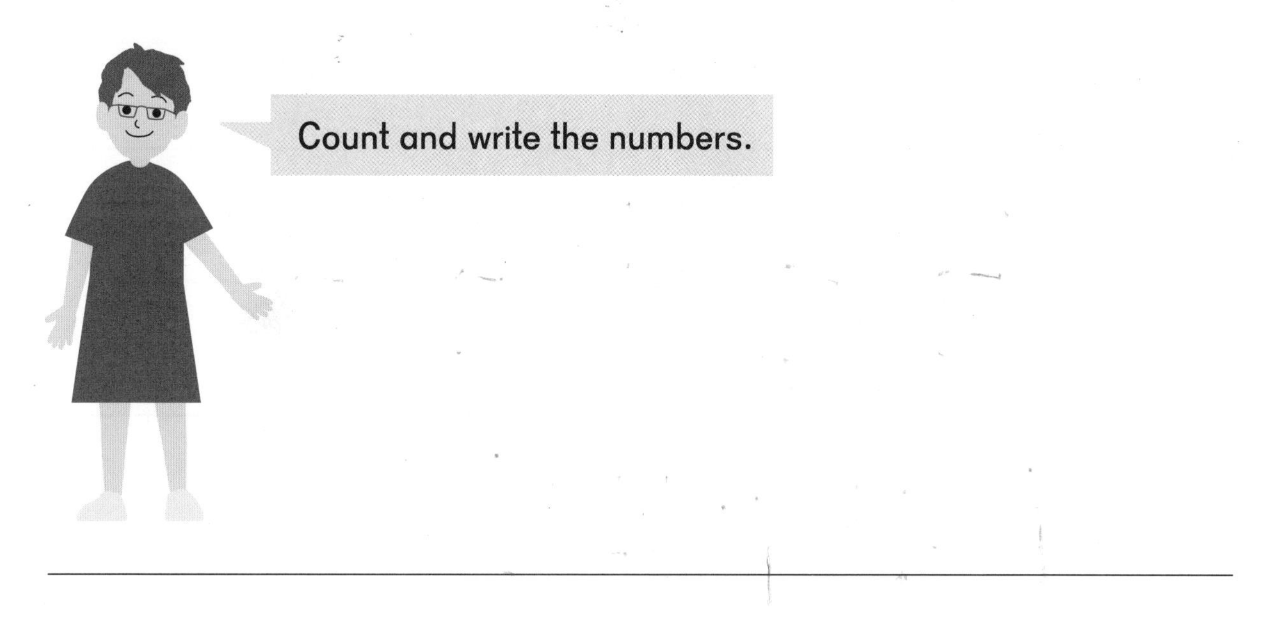

Objective: Write the numerals 1 to 4 to match a set containing that number of objects.

Exercise 8 • page 31

Lesson 9
Trace and Write 1 to 5

9

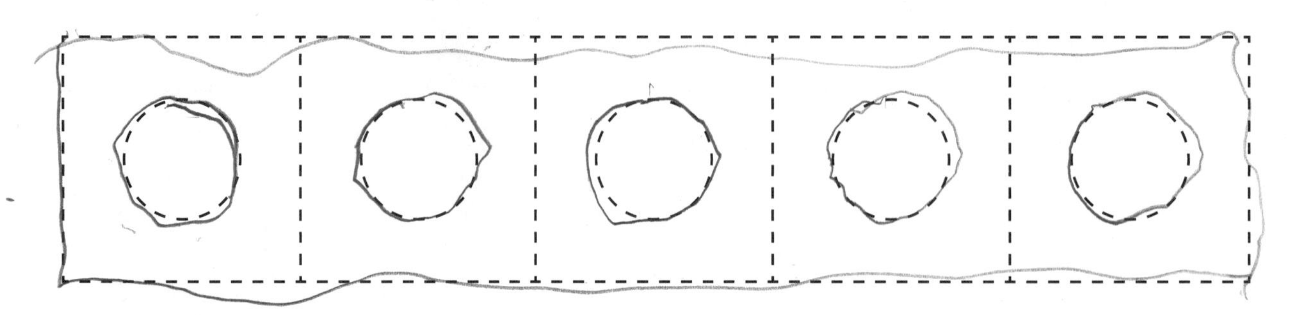

Trace and write.

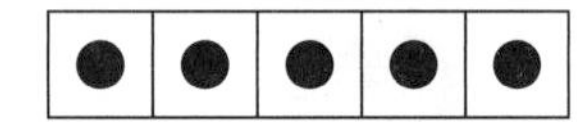

5 5 5 5 5 5 5

5

Objective: Write the numeral 5.

2

Count and write the numbers.

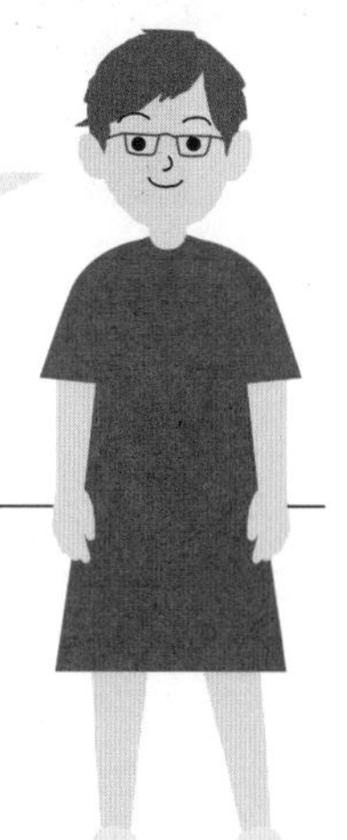

Objective: Write the numerals 1 to 5 to match a set containing that number of objects.

Exercise 9 • page 33

Zero

Objective: Understand that zero represents an empty set.

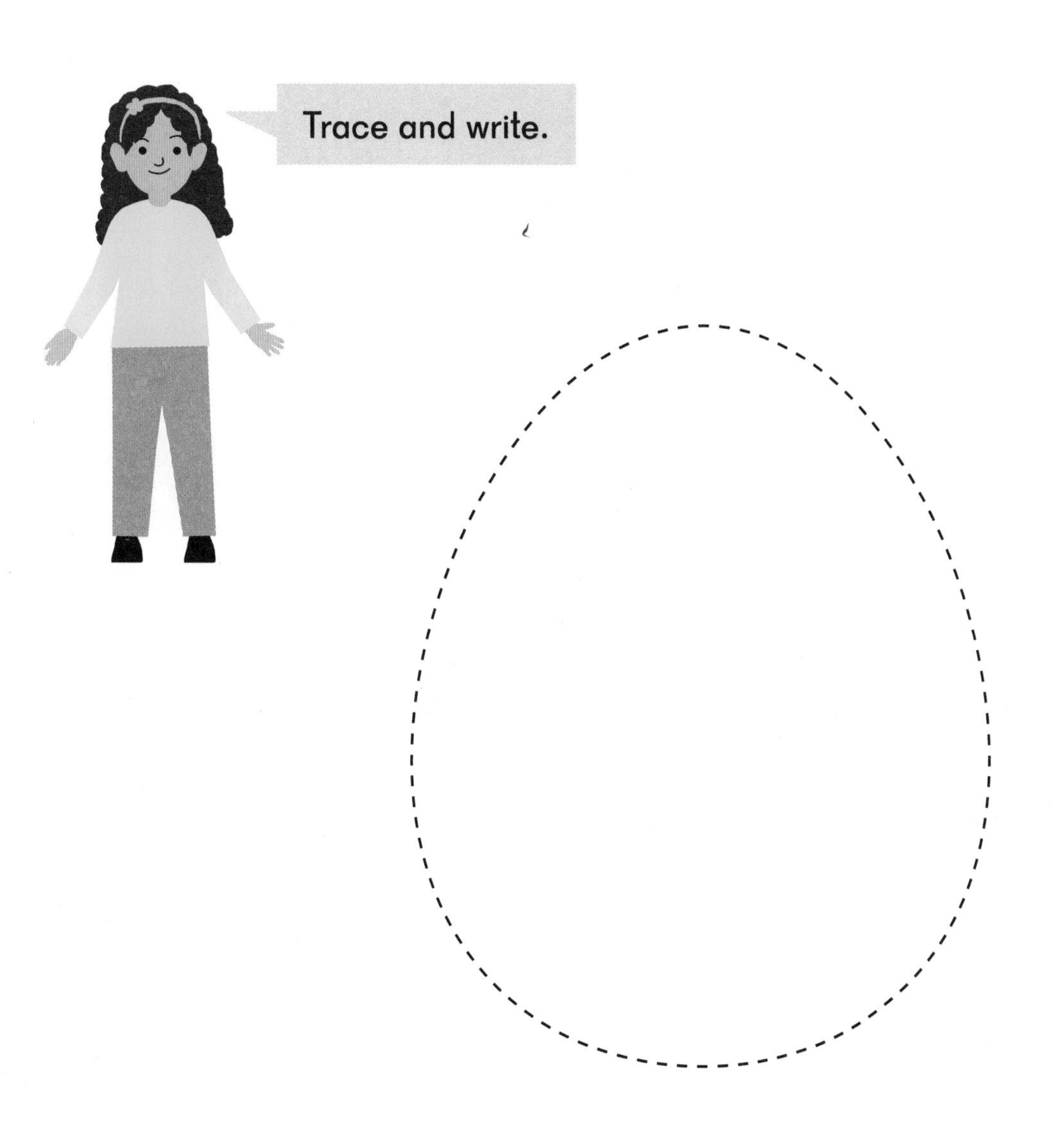

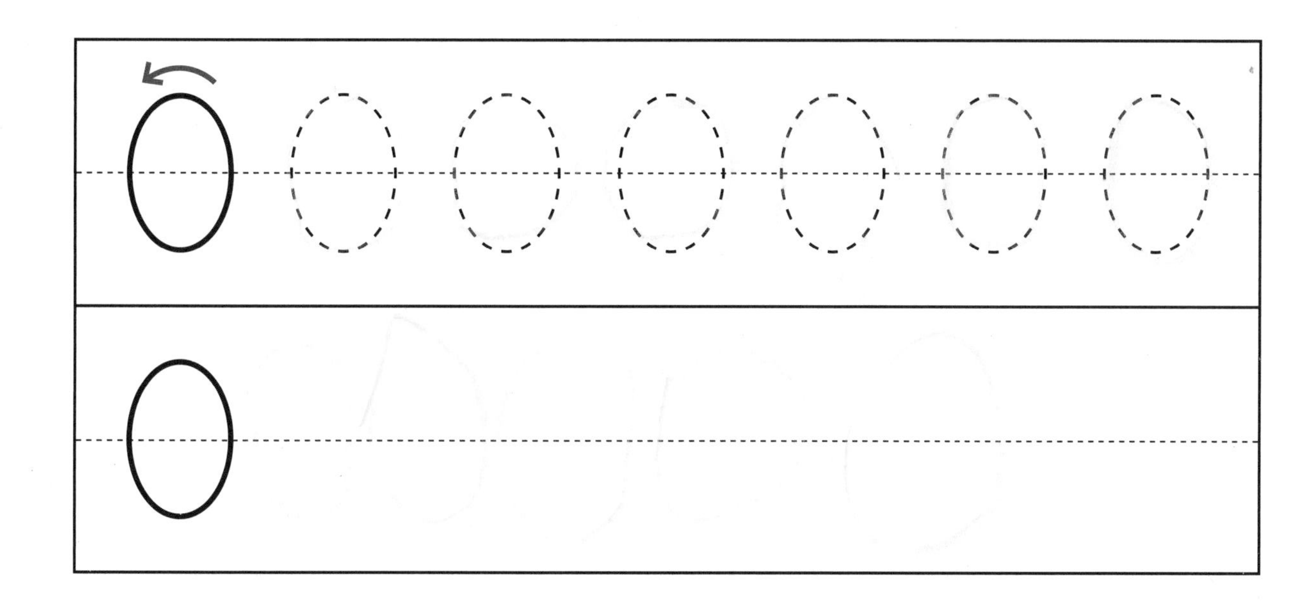

Objective: Write the numeral 0.

Count and write the numbers.

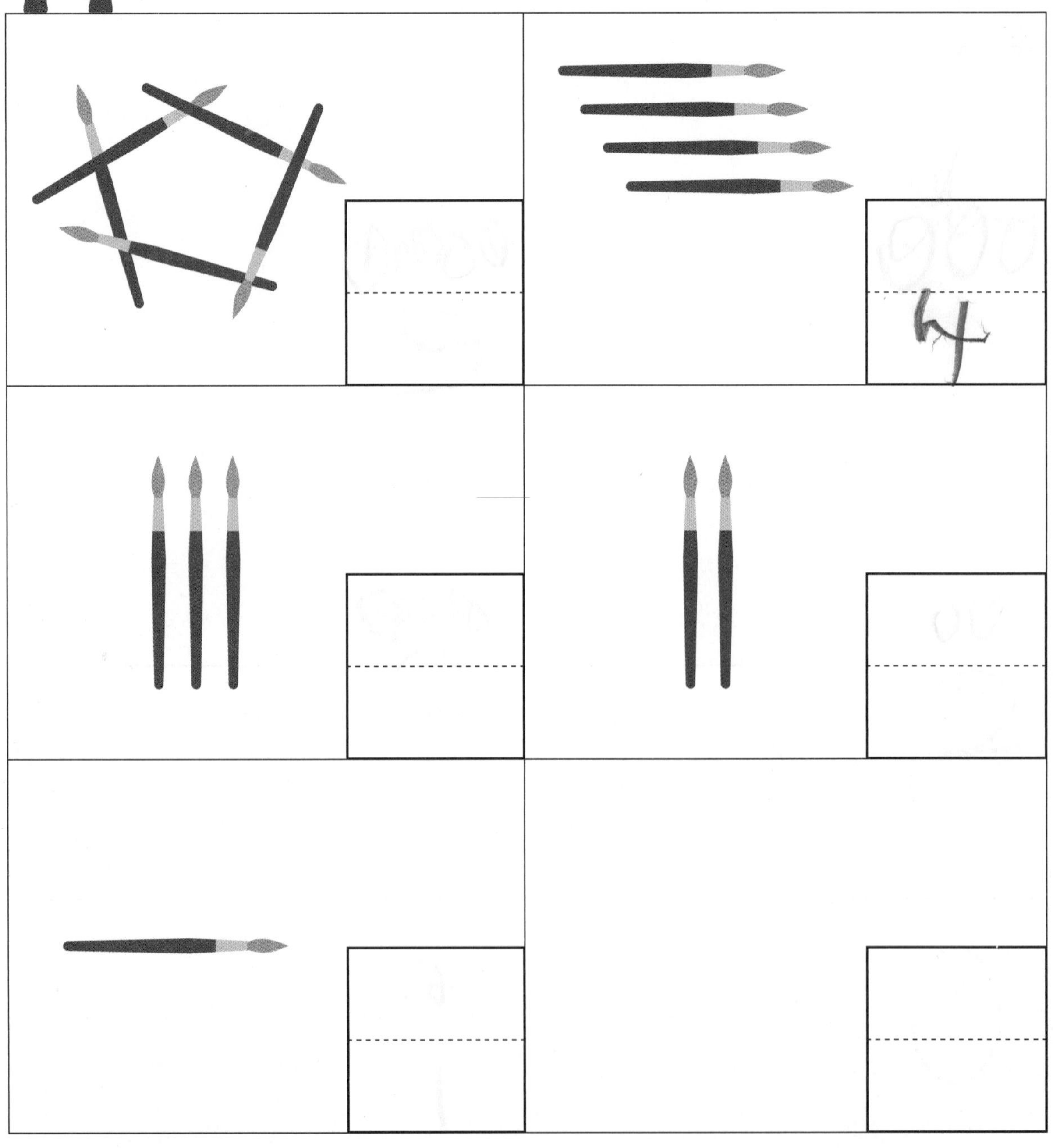

Objective: Write the numerals 0 to 5 to match a set containing that number of objects.

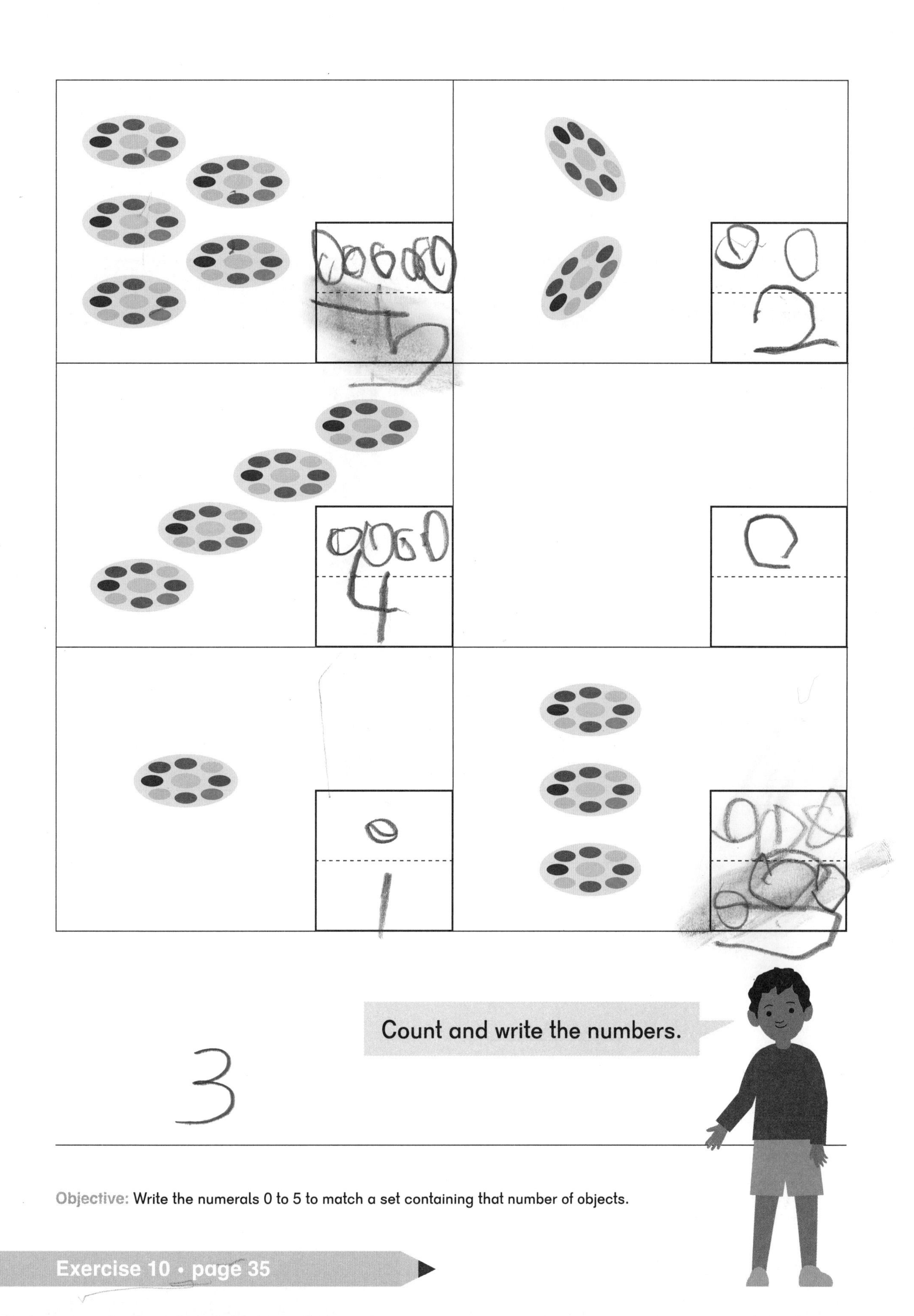

Objective: Write the numerals 0 to 5 to match a set containing that number of objects.

Exercise 10 • page 35

Look and answer the questions.

Favorite Fruits

Objective: Interpret data by looking at a picture graph.

Look and answer the questions.

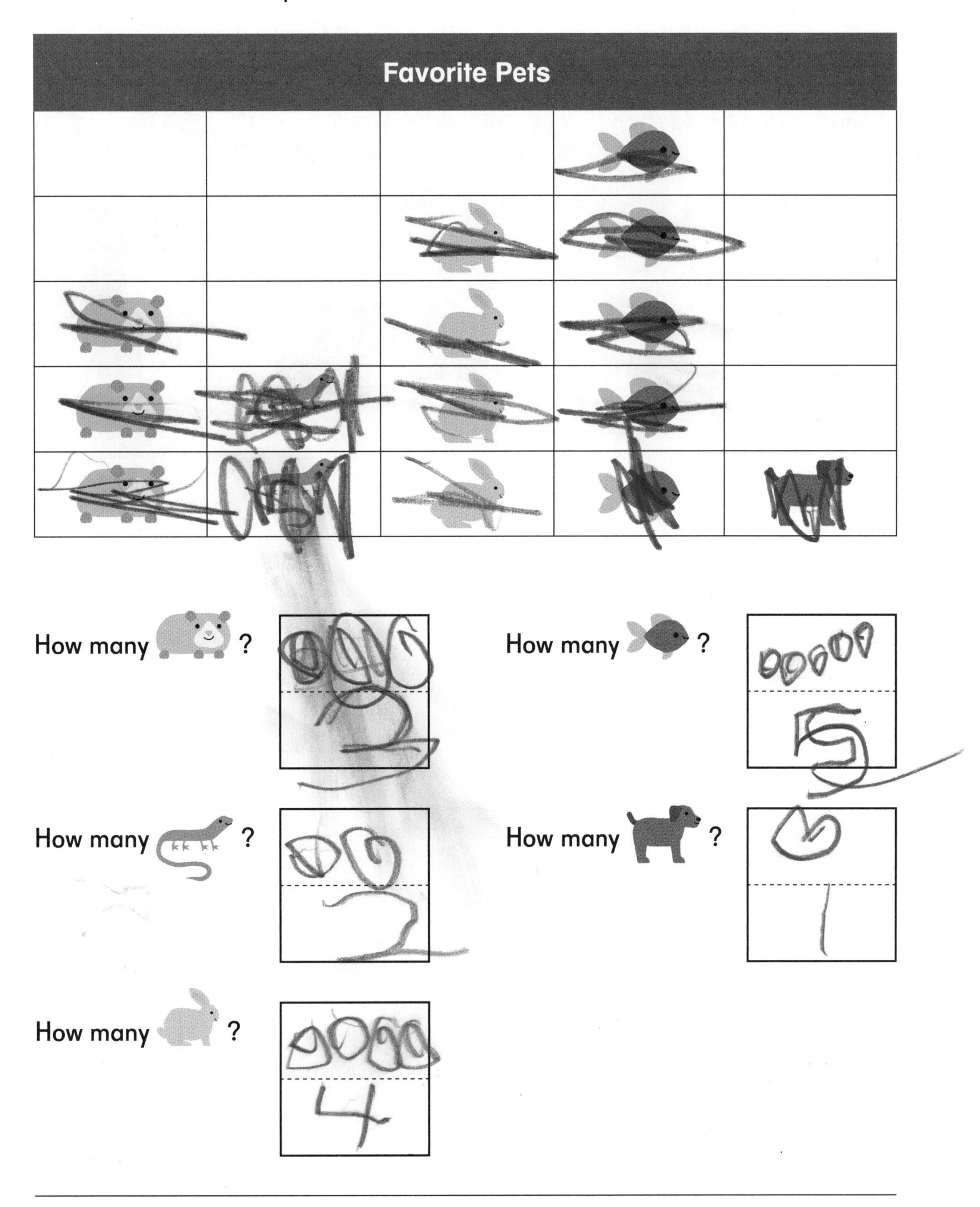

Objective: Interpret data by looking at a picture graph.

Color the boxes to show how many.

Graph Name:				

Objective: Represent data on a graph.

Exercise 11 • page 37

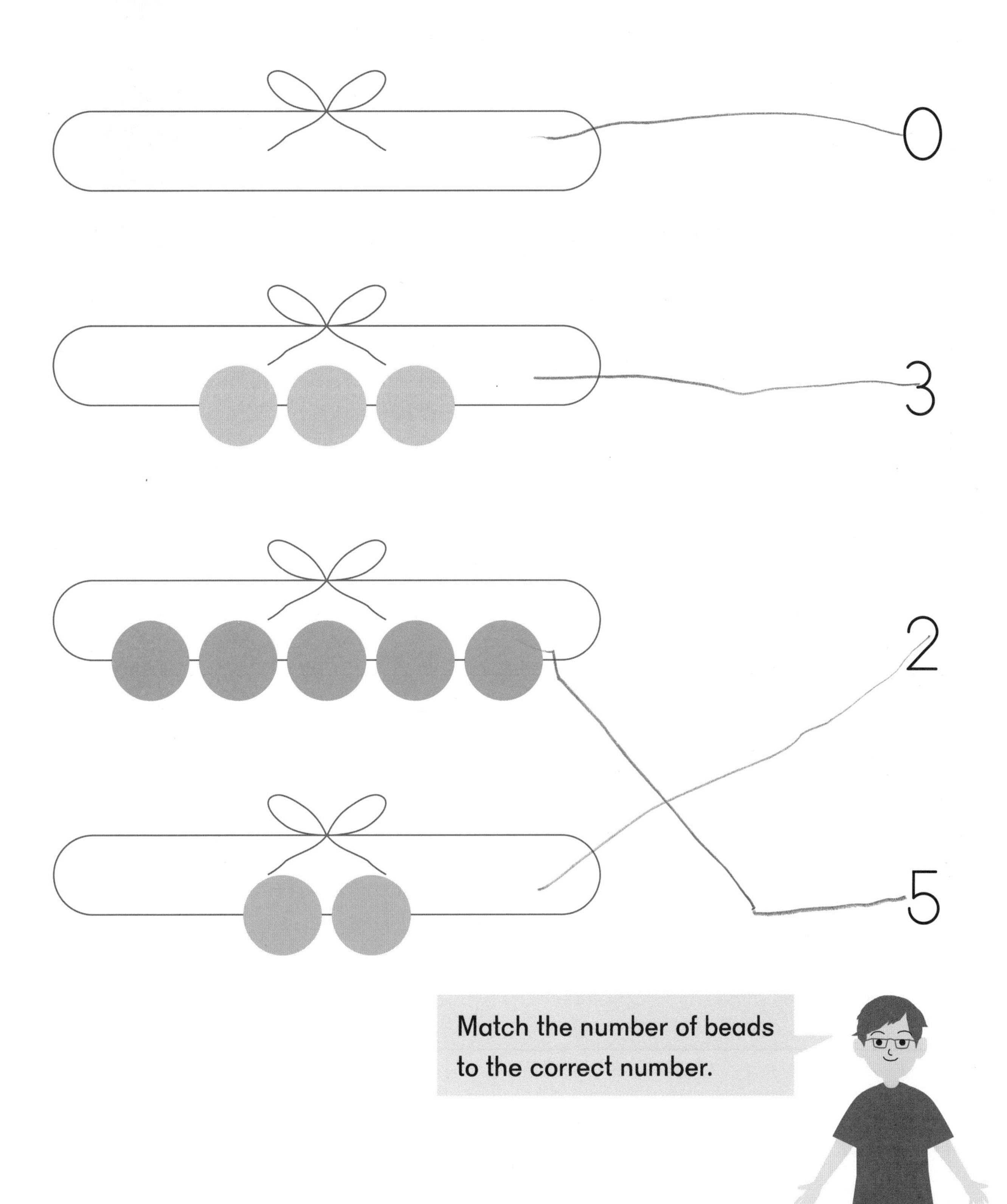

Objective: Practice.

Draw the correct number of circles next to each number.

3
0
5
2
1
4

Objective: Practice.

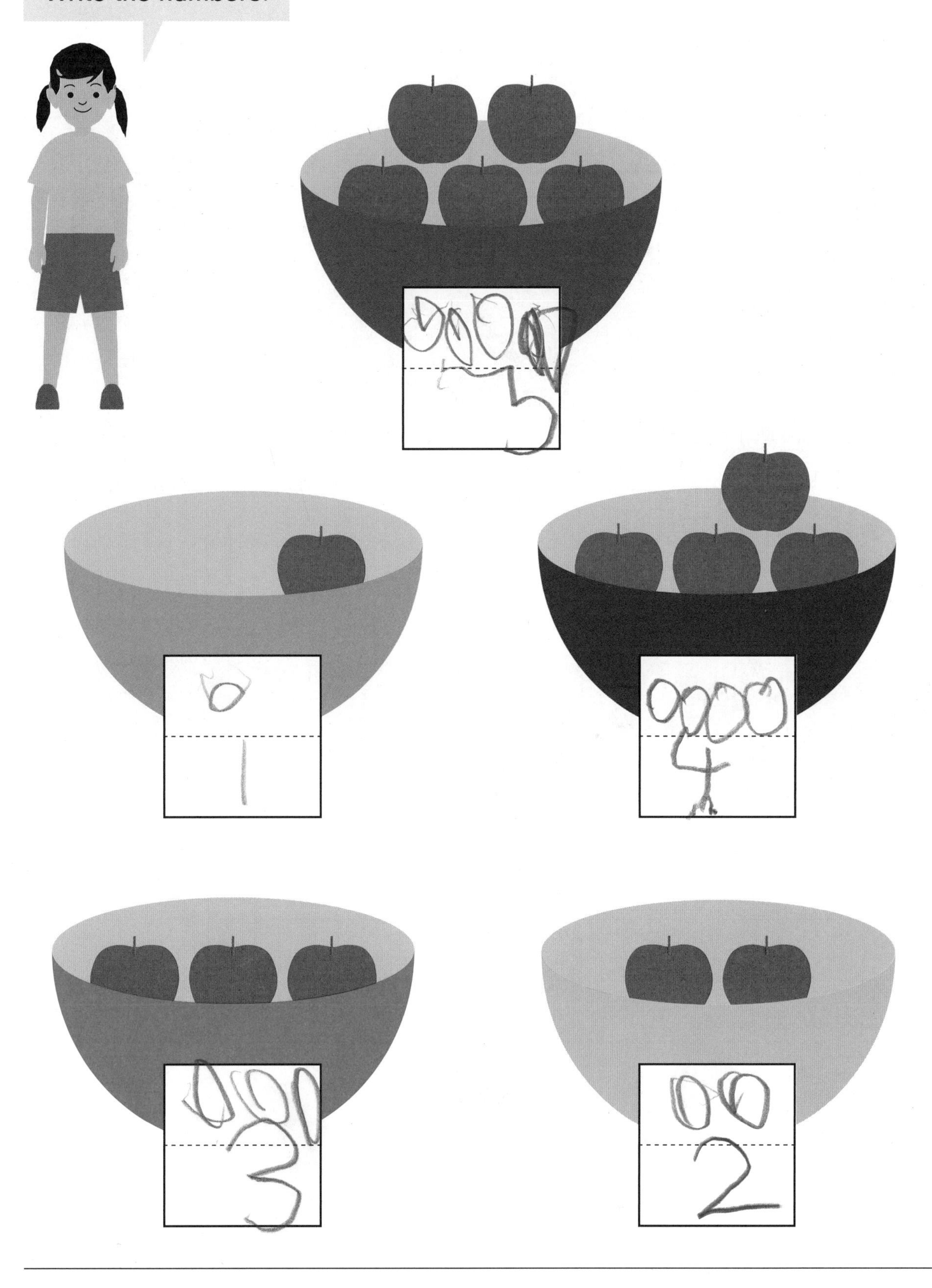

Objective: Practice.

Color in the picture graph.

Farm Animals				

Objective: Practice.

Exercise 12 • page 39

Chapter 3

Numbers to 10

Lesson 1
Count 1 to 10

1

Objective: Count 1 to 10 by rote and using one-to-one correspondence.

Lesson 2
Count Up to 7 Things

2

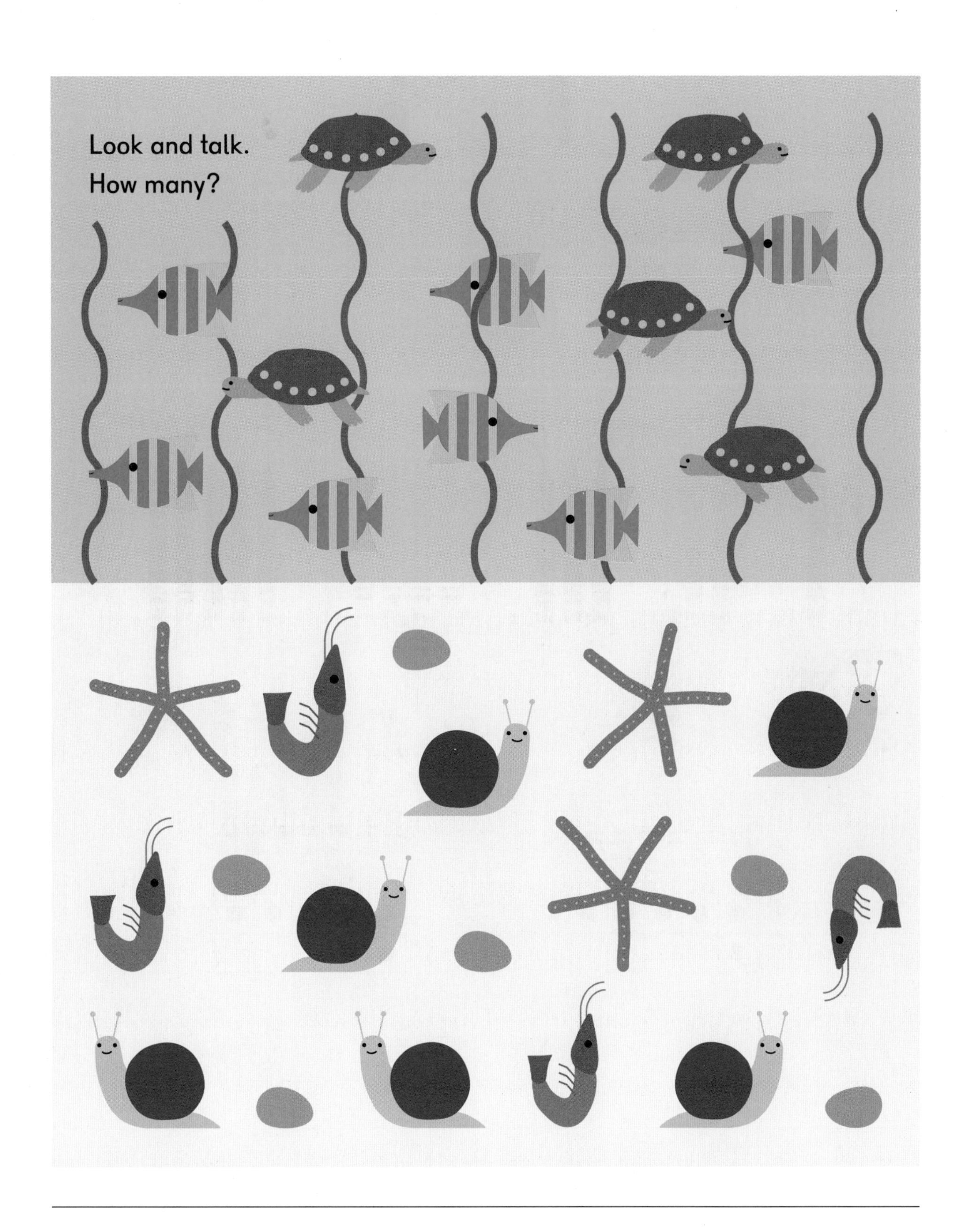

Objective: Count up to 7 objects.

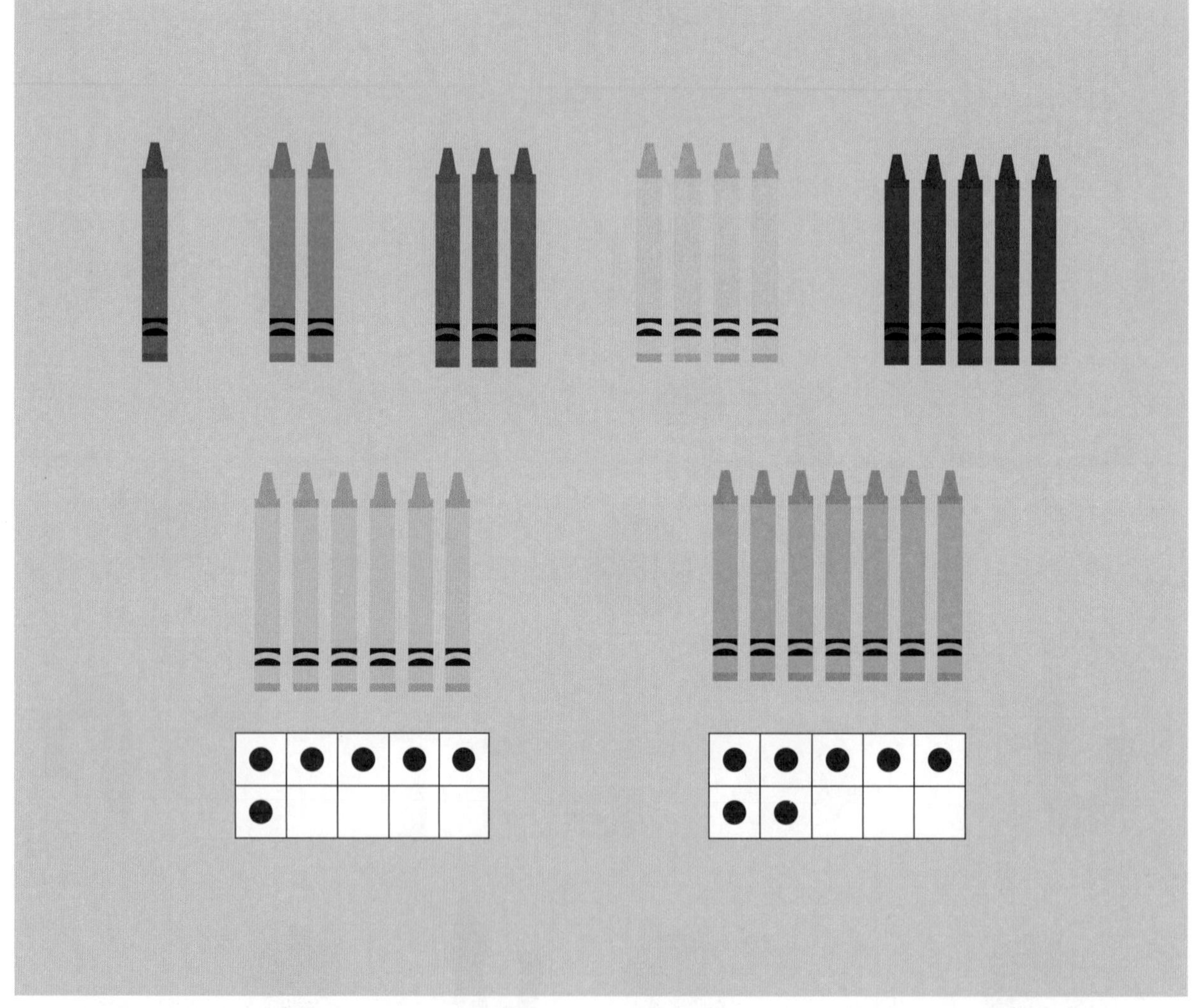

Objective: Count up to 7 objects in a set and say how many are in the set.

Exercise 1 • page 43

Lesson 3
Count Up to 9 Things

3

Look and talk.
How many?

Objective: Count up to 9 objects and represent them on a ten-frame card.

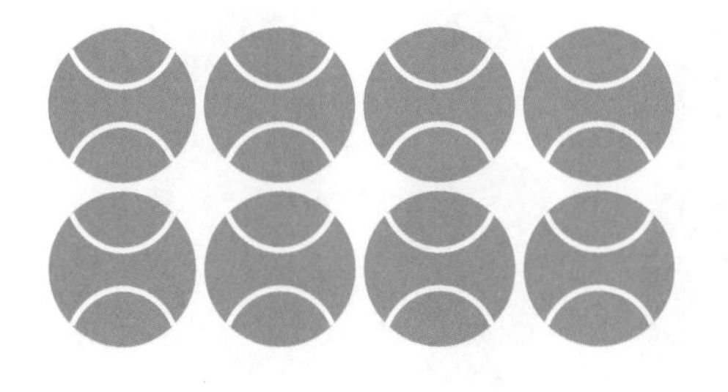

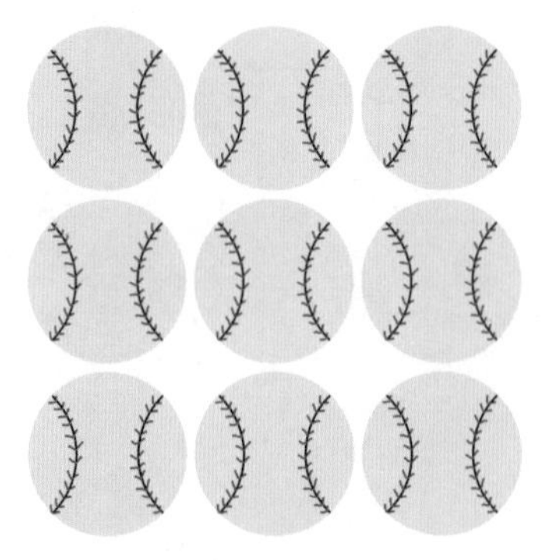

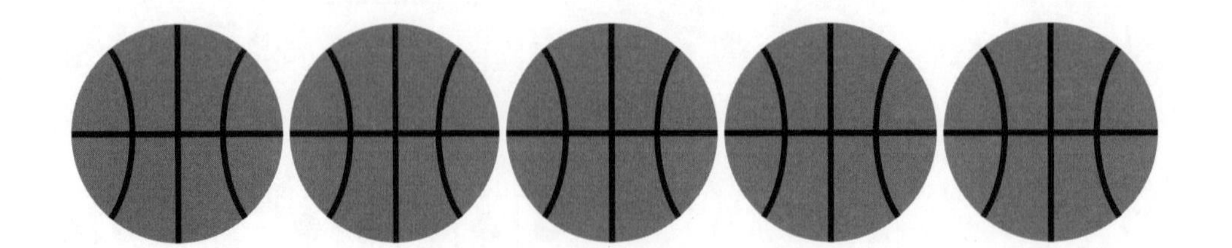

Objective: Count up to 9 objects in a set and say how many objects are in the set.

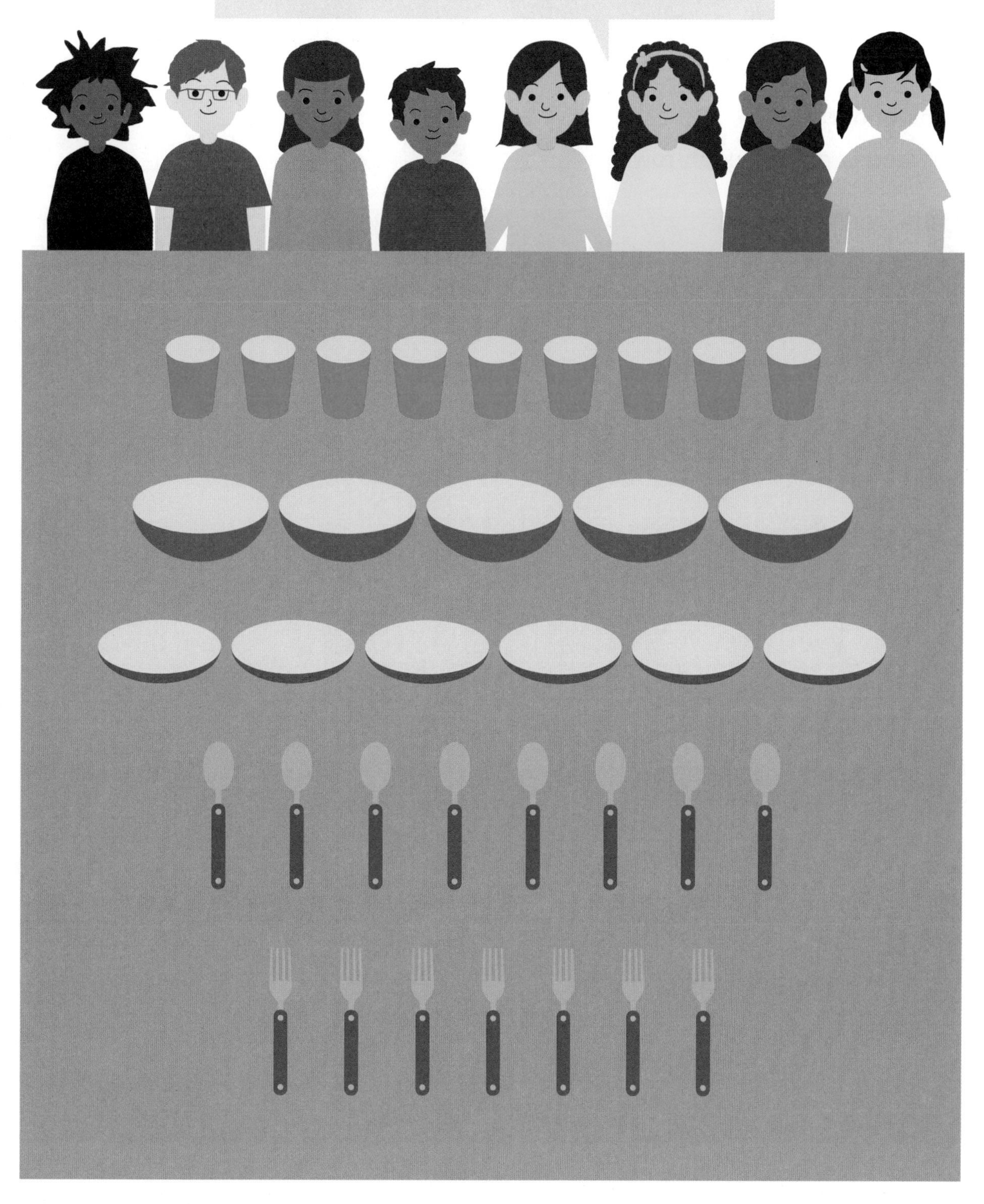

Objective: Count up to 9 objects and represent the number on a ten-frame card.

Exercise 2 • page 45

Lesson 4
Count Up to 10 Things — Part 1

Look and talk.

How many ants in each row?

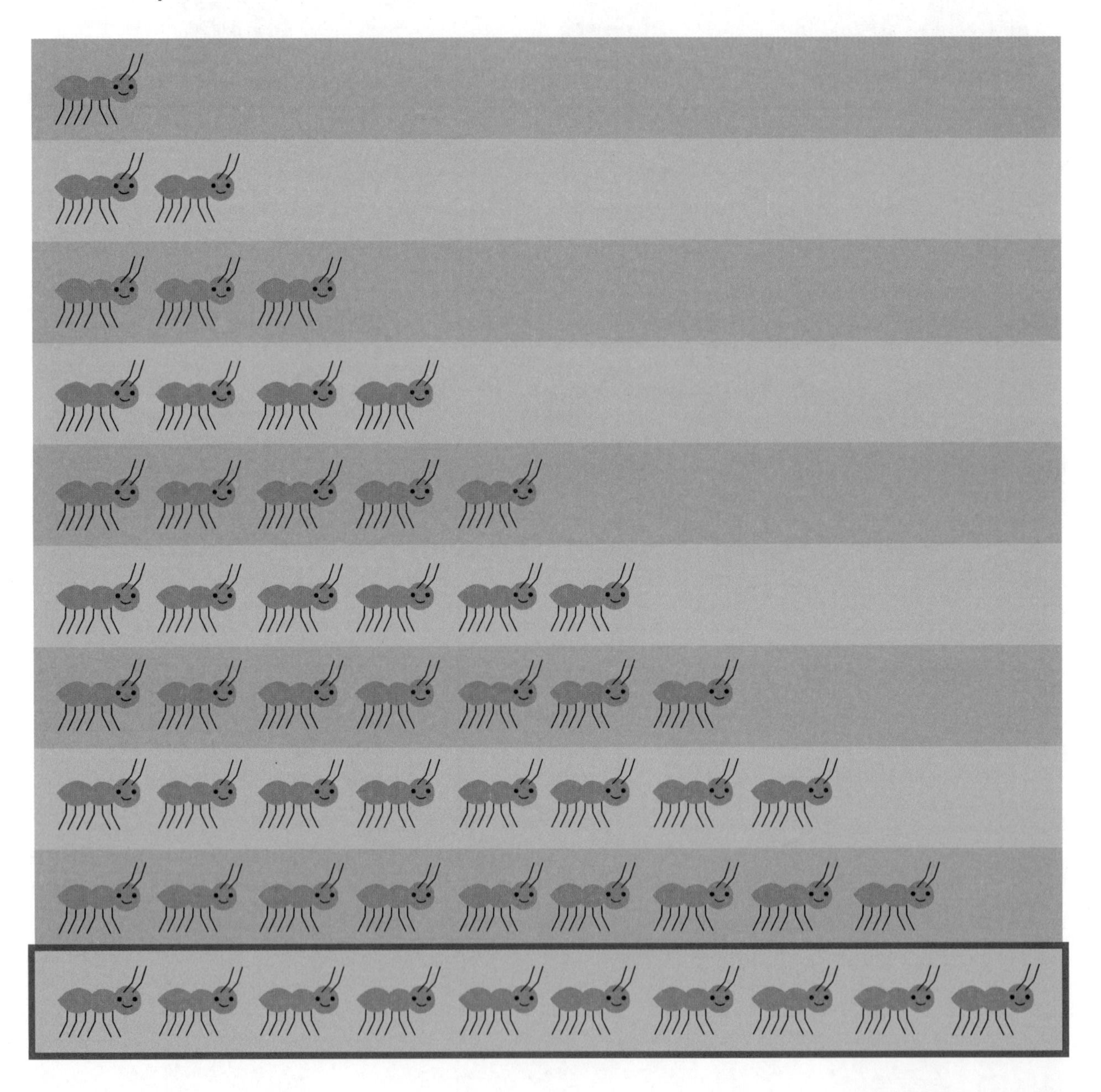

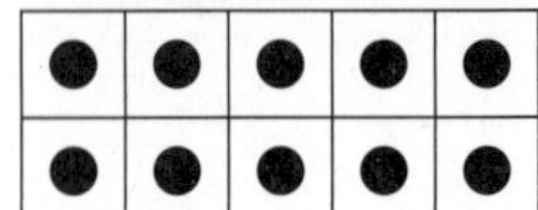

Objective: Count up to 10 objects in a set and represent the number on a ten-frame card.

Objective: Count up to 10 objects in a set and say how many are in the set.

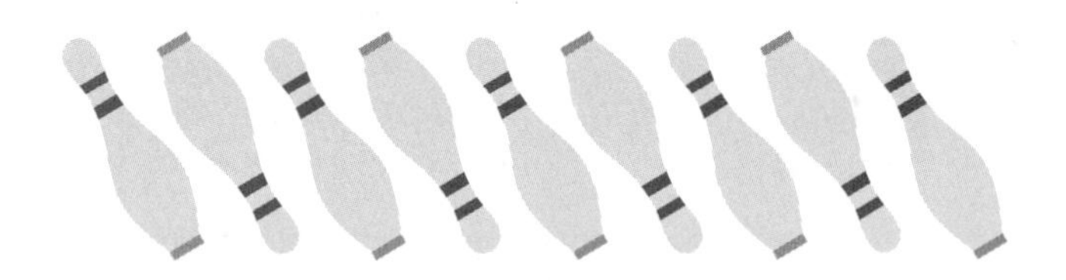

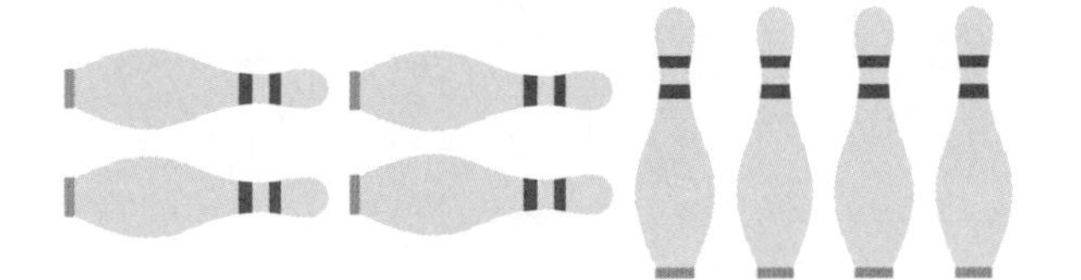

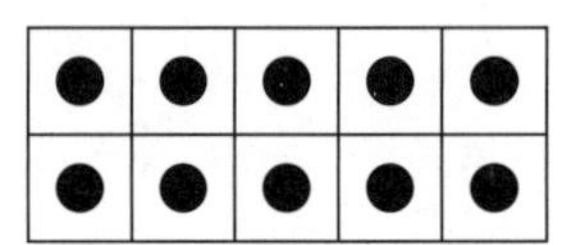

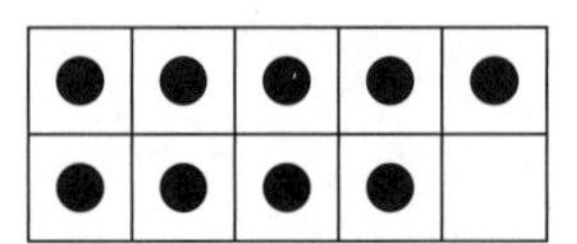

Match the bowling pins to the correct ten-frame card.

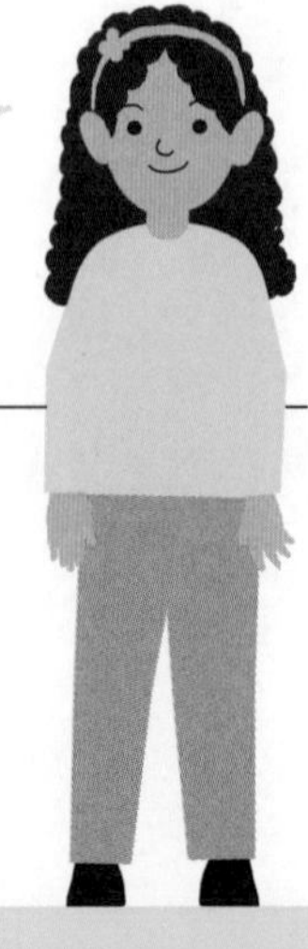

Lesson 5
Count Up to 10 Things – Part 2

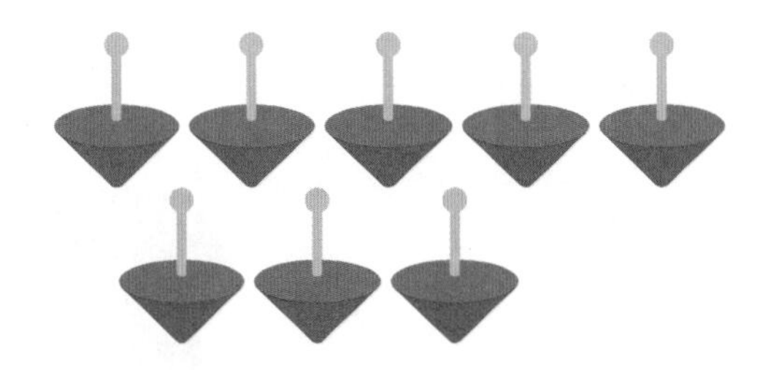

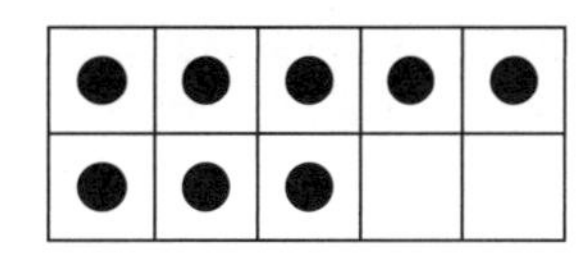

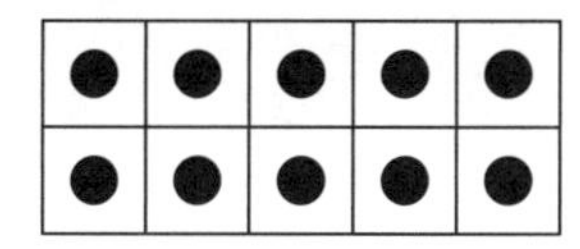

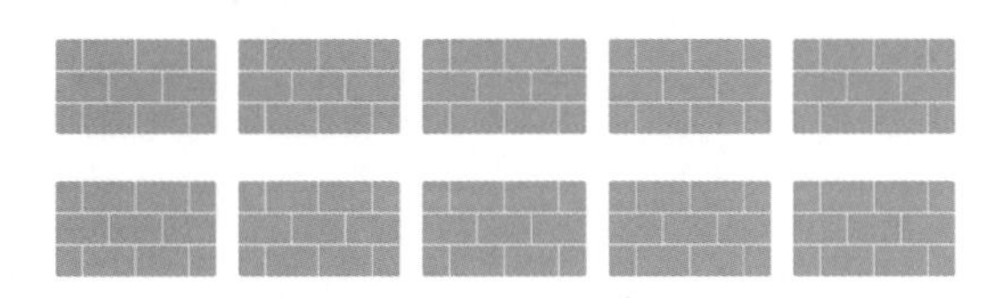

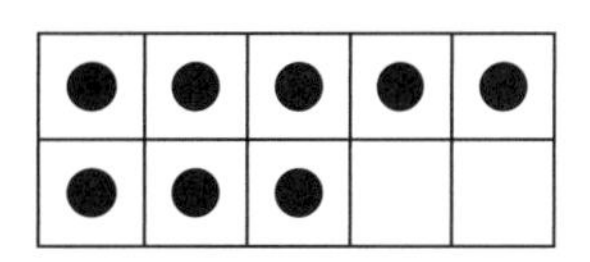

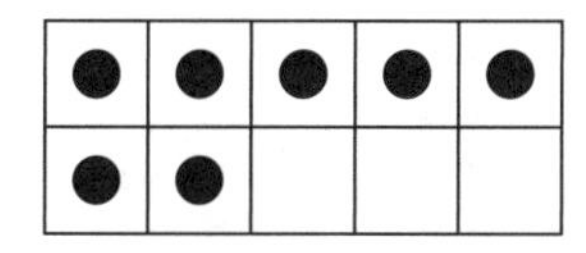

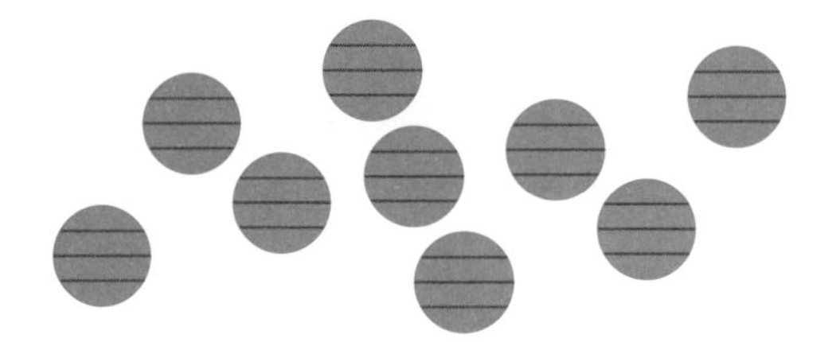

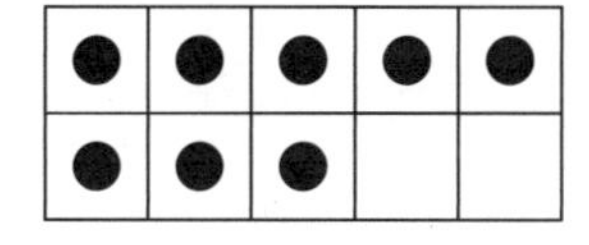

Count the things and circle the ten-frame card that matches.

Objective: Say how many, and match with a ten-frame card.

Exercise 4 • page 49

Lesson 6
Recognize the Numbers 6 to 10

Match the vegetables to the correct number.

Objective: Recognize the numerals 6 to 10 and match a set of objects to the correct numeral.

10

7

6

9

Color in each ten-frame card to show the correct number.

Objective: Recognize the numerals 6 to 10 and represent the number on a ten-frame card.

Exercise 5 • page 53

Lesson 7
Write the Numbers 6 and 7

7

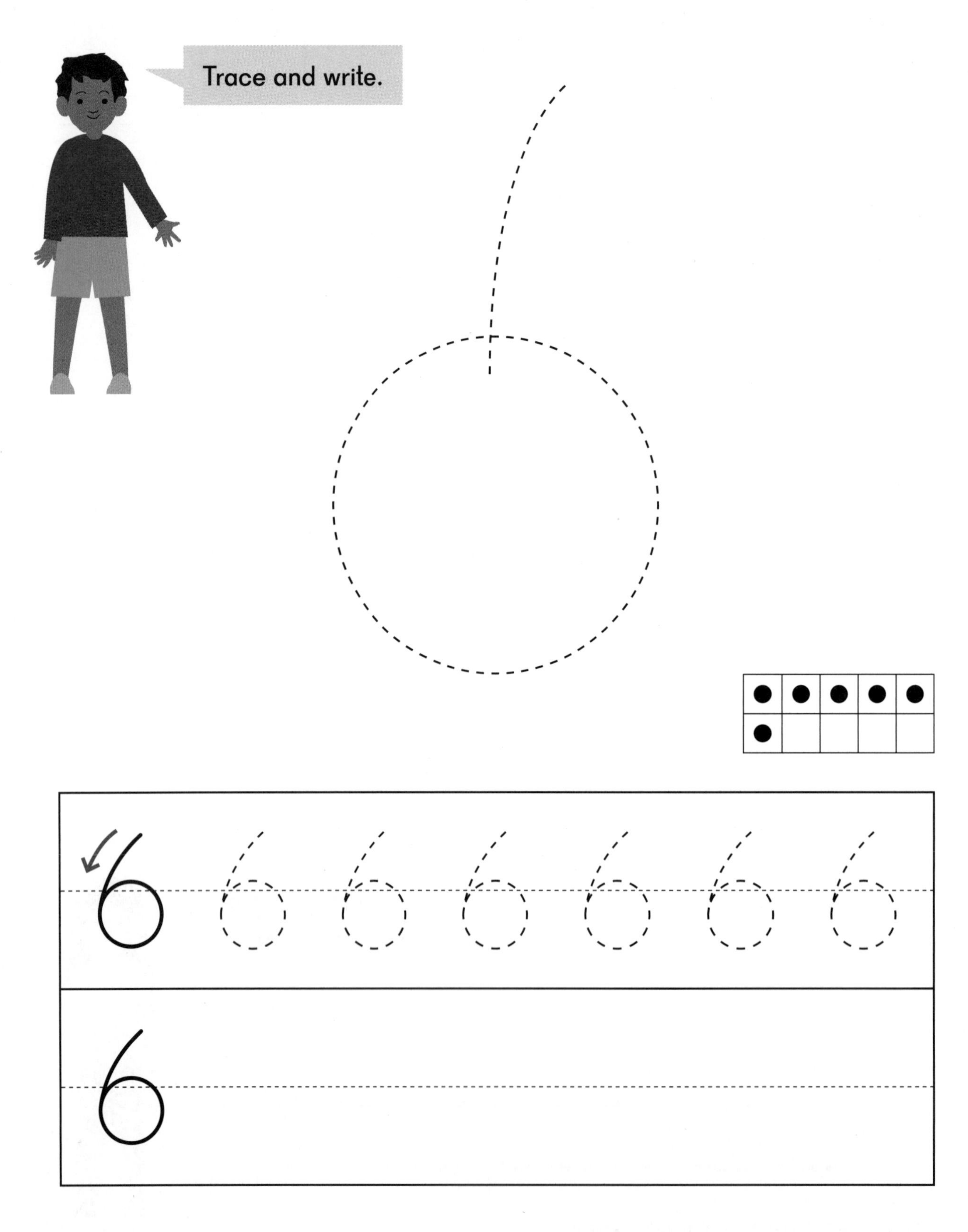

Objective: Write the numeral 6.

Trace and write.

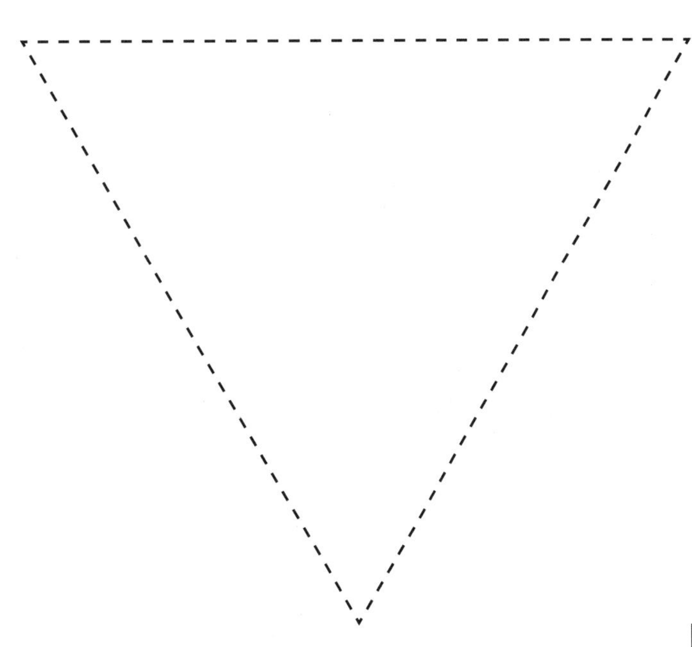

●	●	●	●	●
●	●			

7 7 7 7 7 7 7

7

Objective: Write the numeral 7.

Exercise 6 • page 55

Lesson 8
Write the Numbers 8, 9, and 10

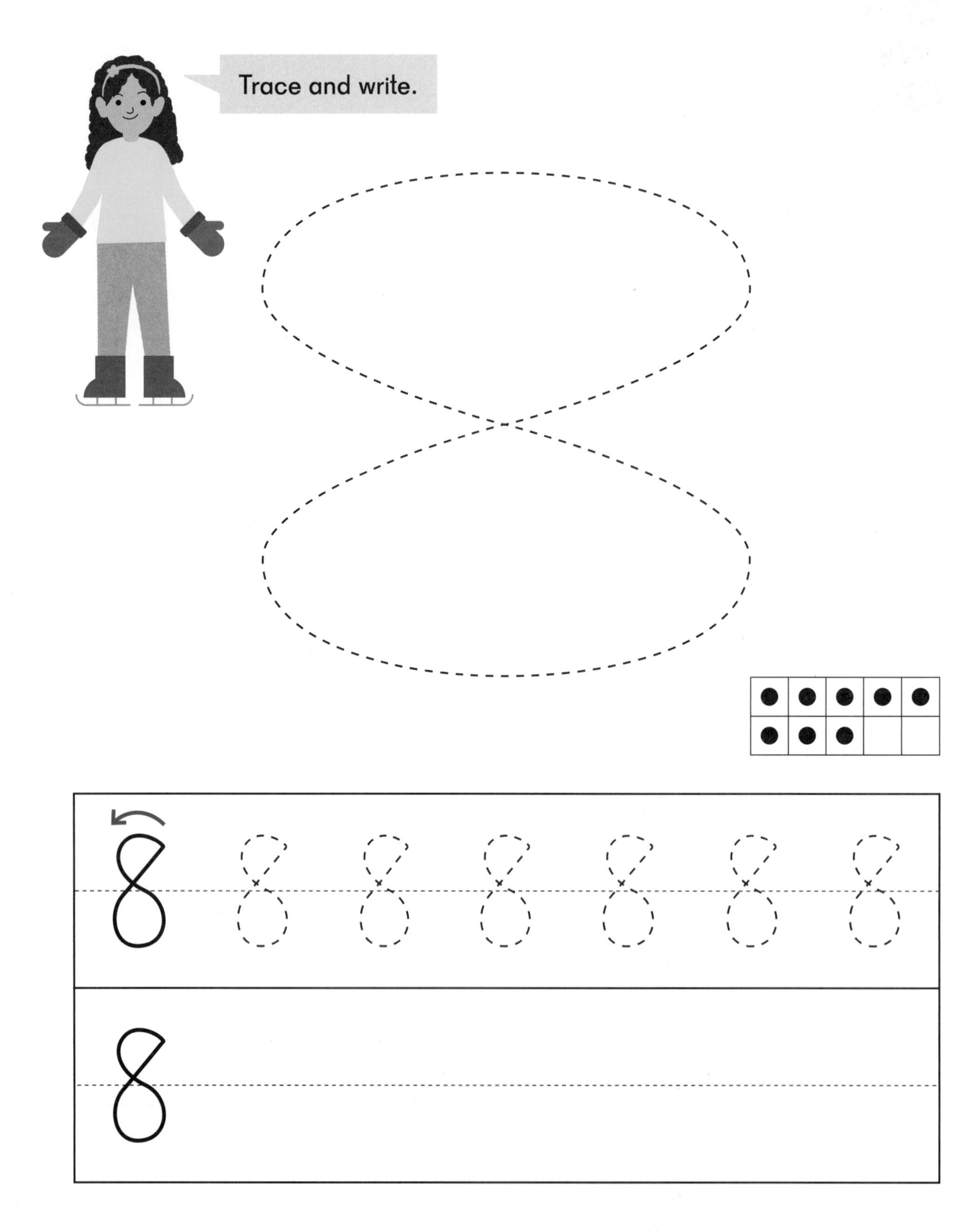

Objective: Write the numeral 8.

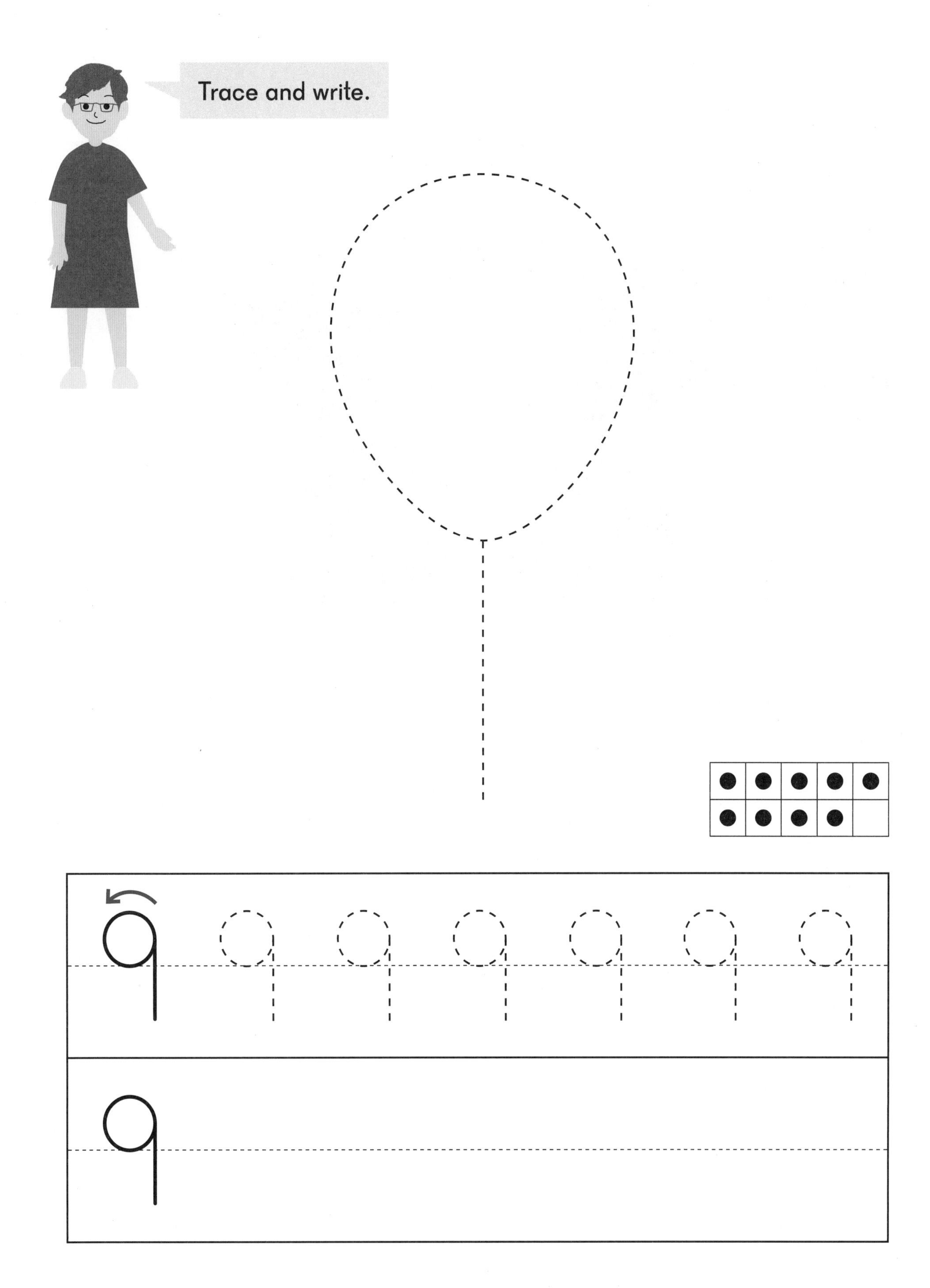

Objective: Write the numeral 9.

Objective: Write the numeral 10.

Exercise 7 • page 57

Lesson 9
Write the Numbers 6 to 10

9

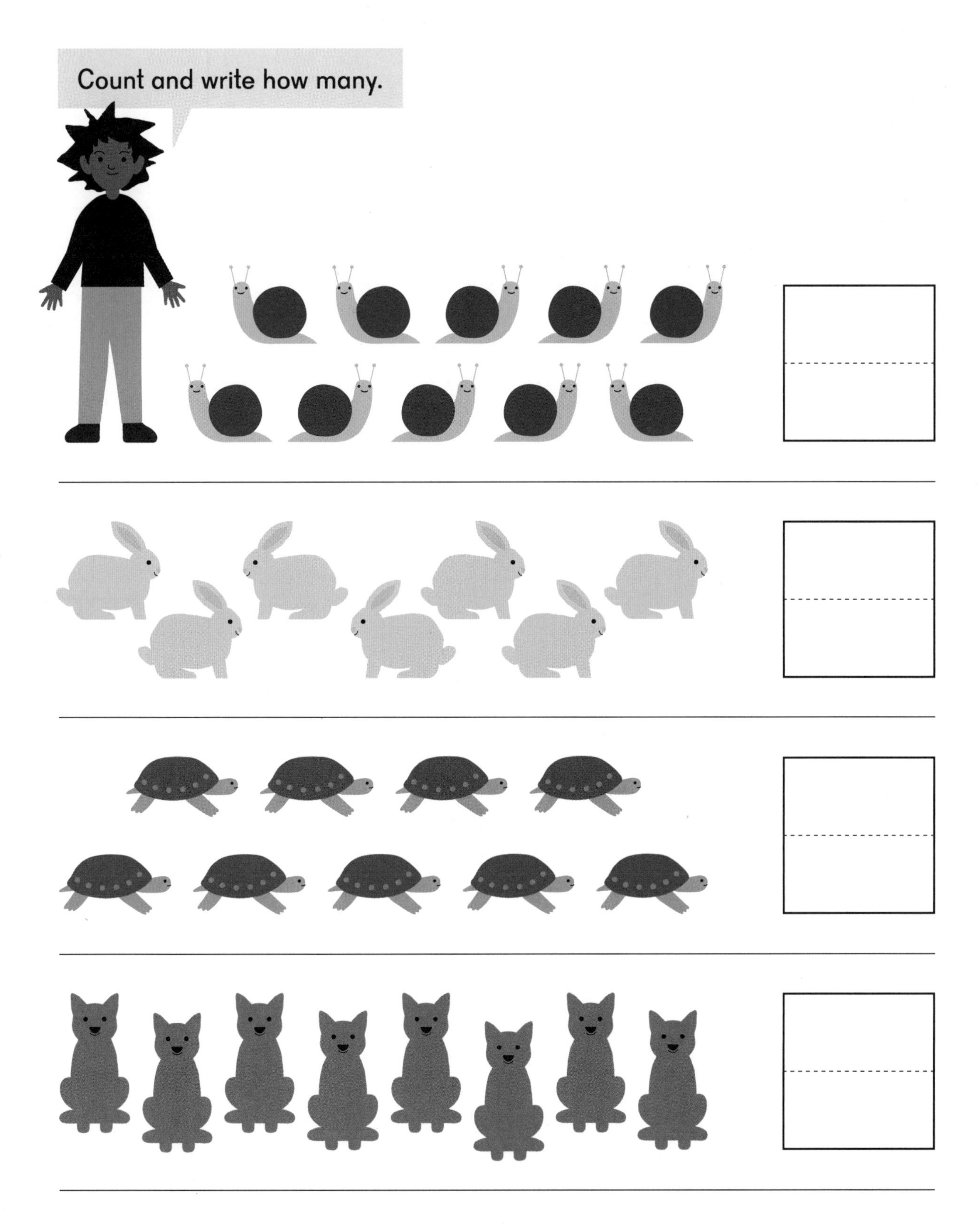

Objective: Write the numerals 6 to 10 to represent the number of objects in a set.

Write the numbers.

Objective: Write the numerals 6 to 10 to represent the number represented on a ten-frame card.

Exercise 8 • page 61

Lesson 10
Count and Write the Numbers 1 to 10

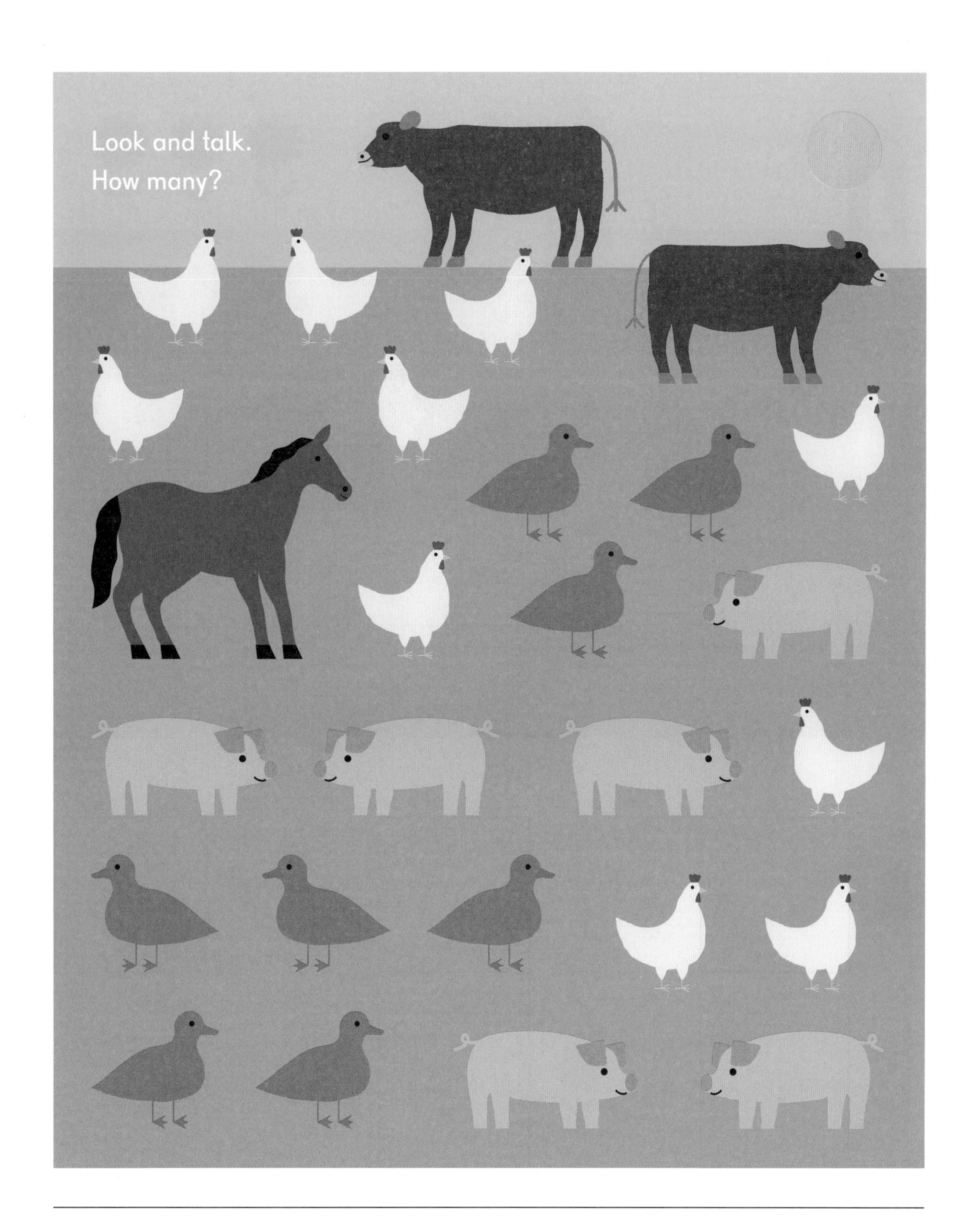

Objective: Count up to 10 objects.

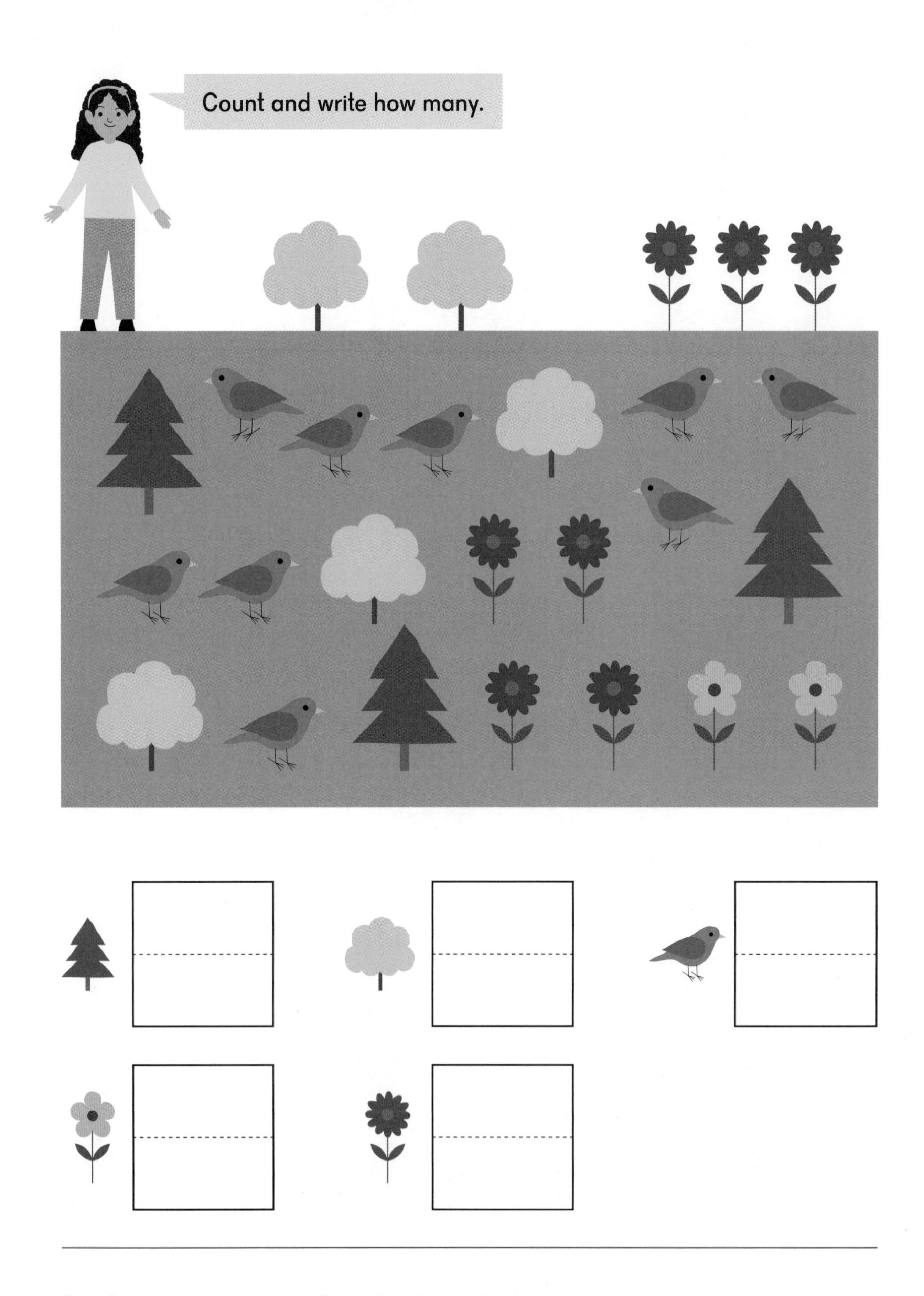

Objective: Write the numerals 1 to 10 to match a set of objects containing that number.

Count and write how many.

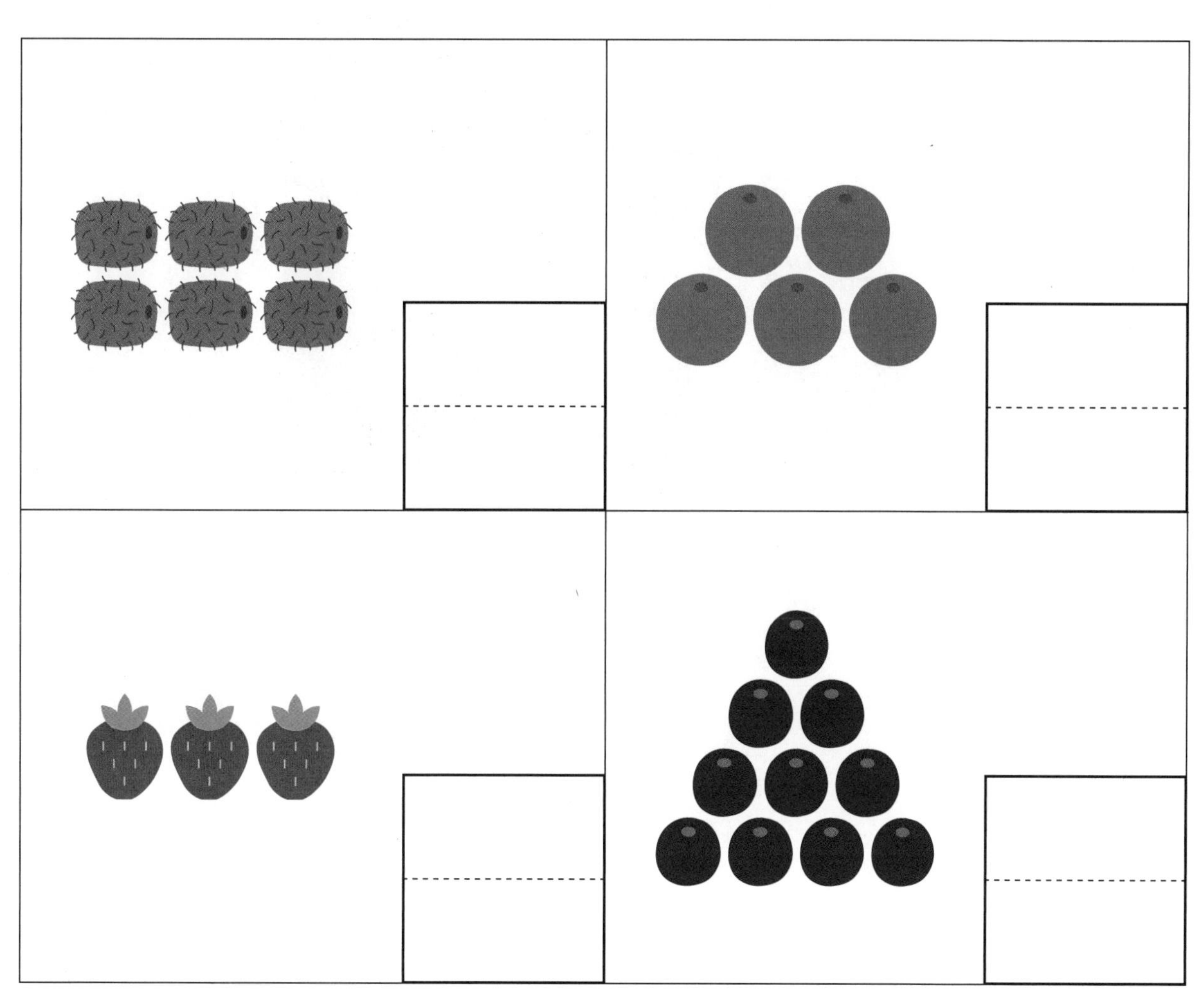

Objective: Write a numeral 1 to 10 to represent the number of objects in a set.

Exercise 9 • page 63

Lesson 11
Ordinal Positions

Objective: Recognize ordinal positions first through fifth from the front.

Color the third crayon from the left red.
Color the fifth crayon from the left green.
Color the seventh crayon from the right brown.

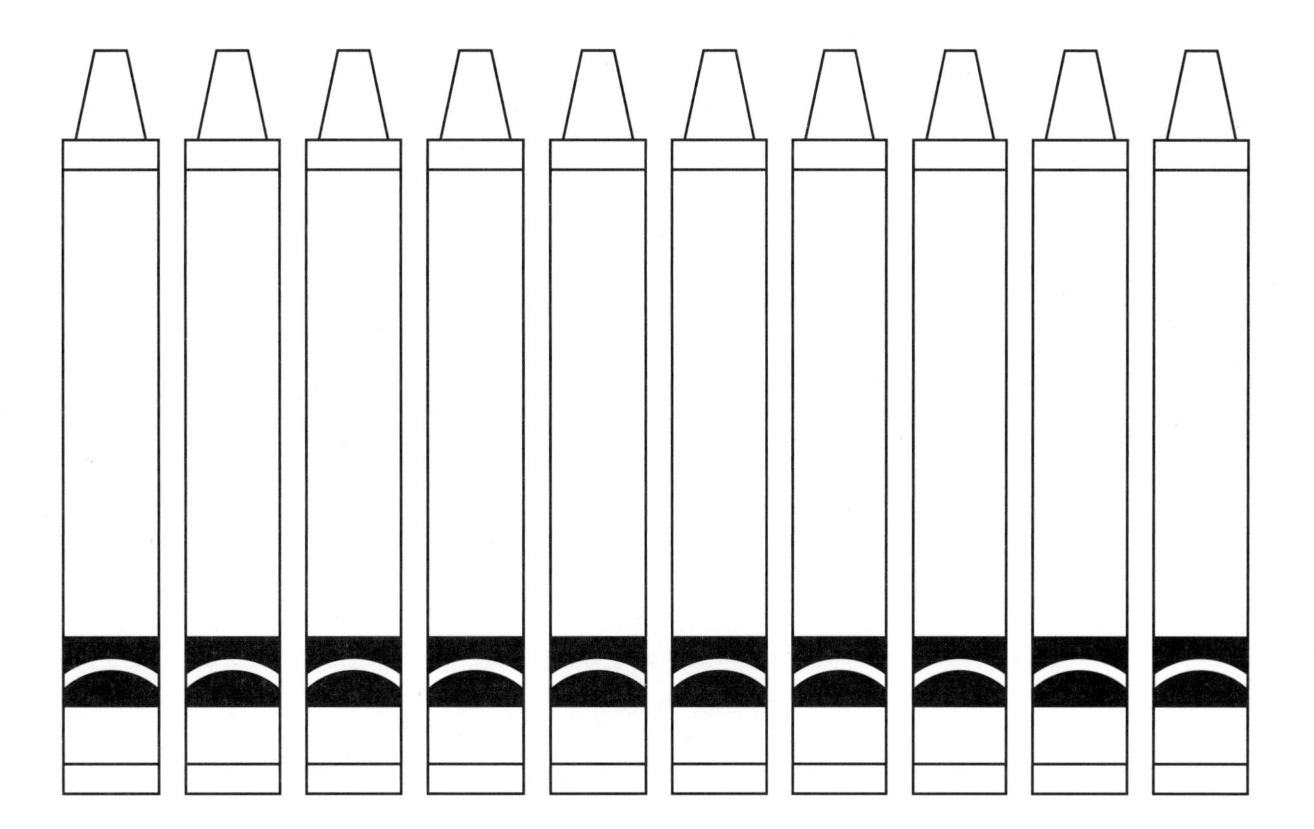

Objective: Identify ordinal positions first through tenth starting from the left or the right.

Cross out the third rock from the bottom.
Circle the eighth rock from the top.

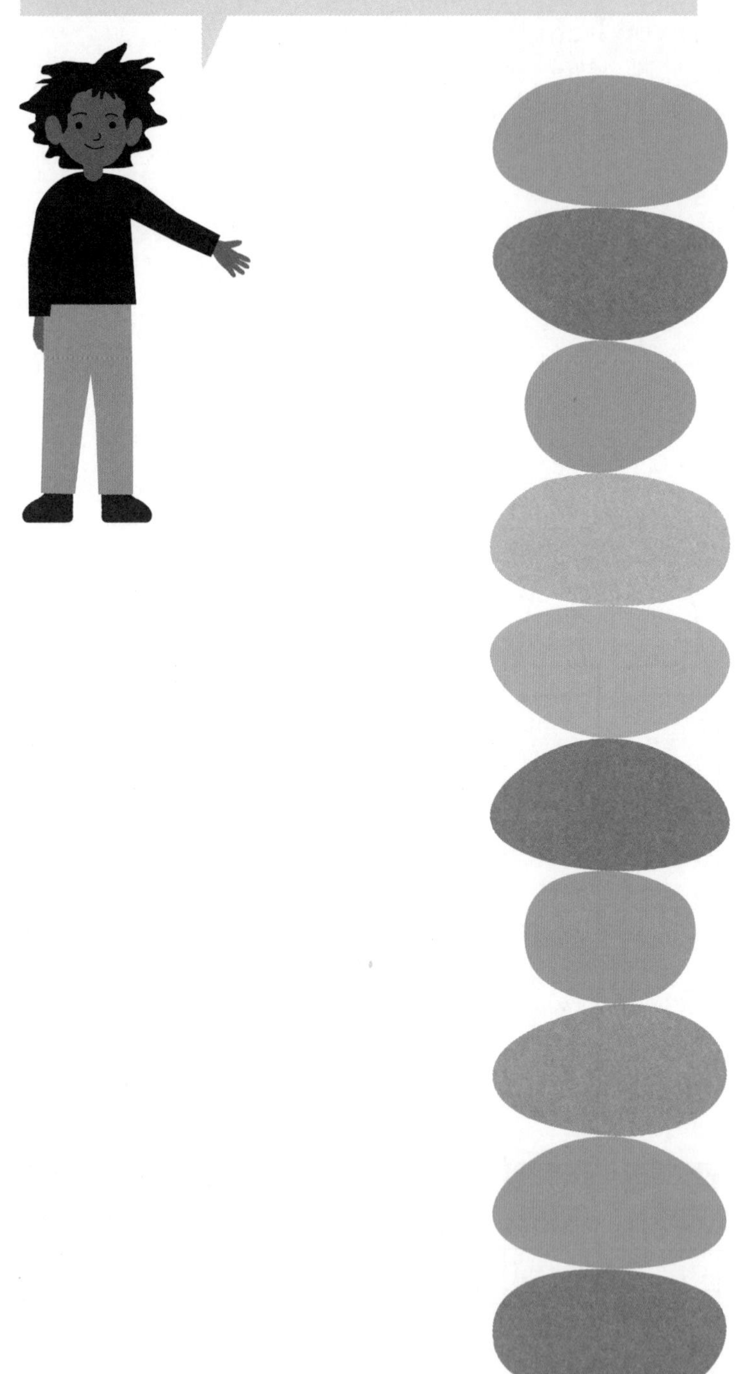

Objective: Identify ordinal positions first through tenth from the top and bottom.

Exercise 10 • page 67

Lesson 12
One More Than

Color the correct number of squares to show one more each time.
Write the number under the squares you colored.
The first three have been done for you.

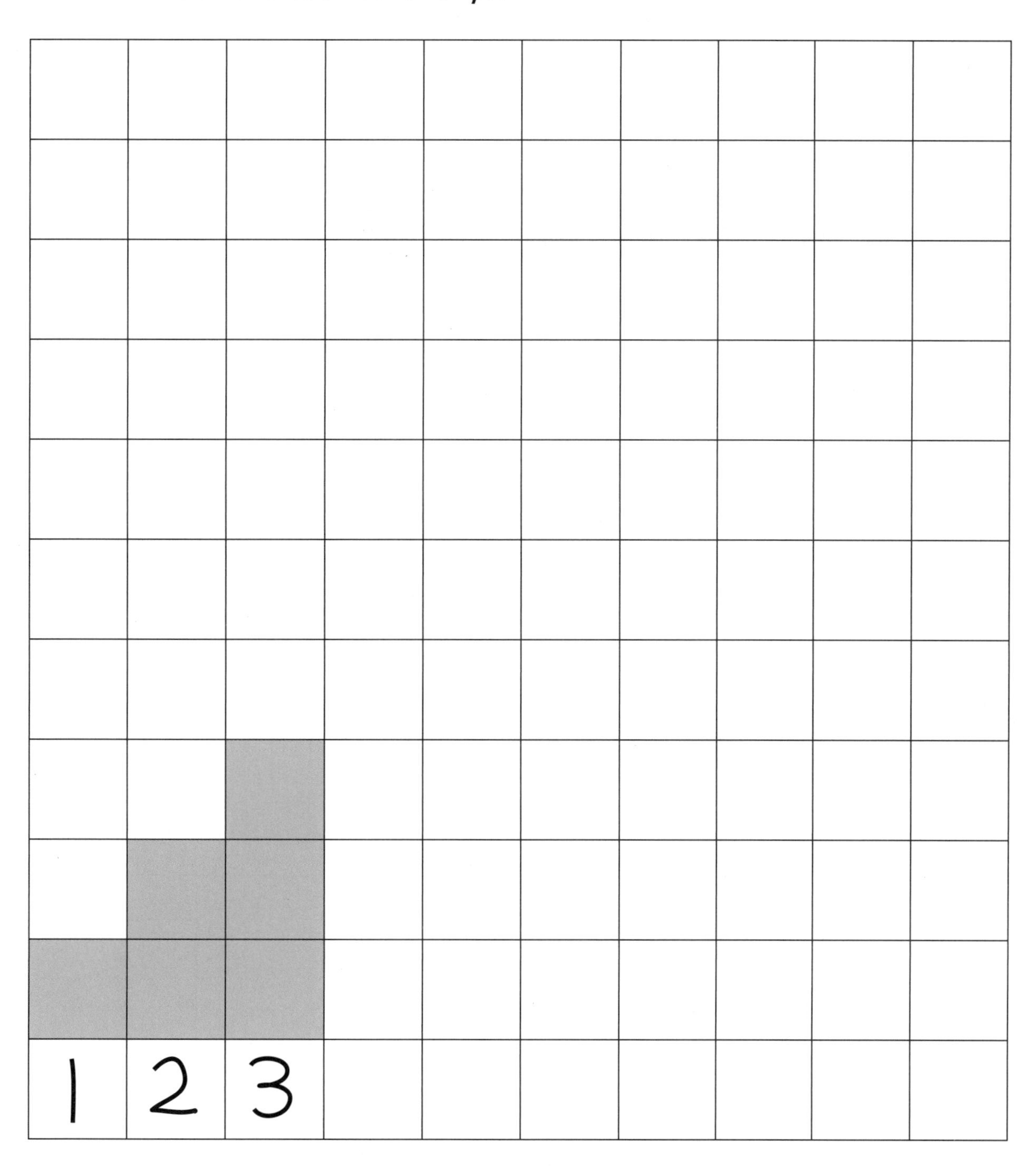

Objective: Recognize that numbers in order increase by 1.

Draw one more and write the number.
The first one is done for you.

1 more

9

Objective: Identify and write the number 1 to 10 that is 1 more than the number of objects in a set.

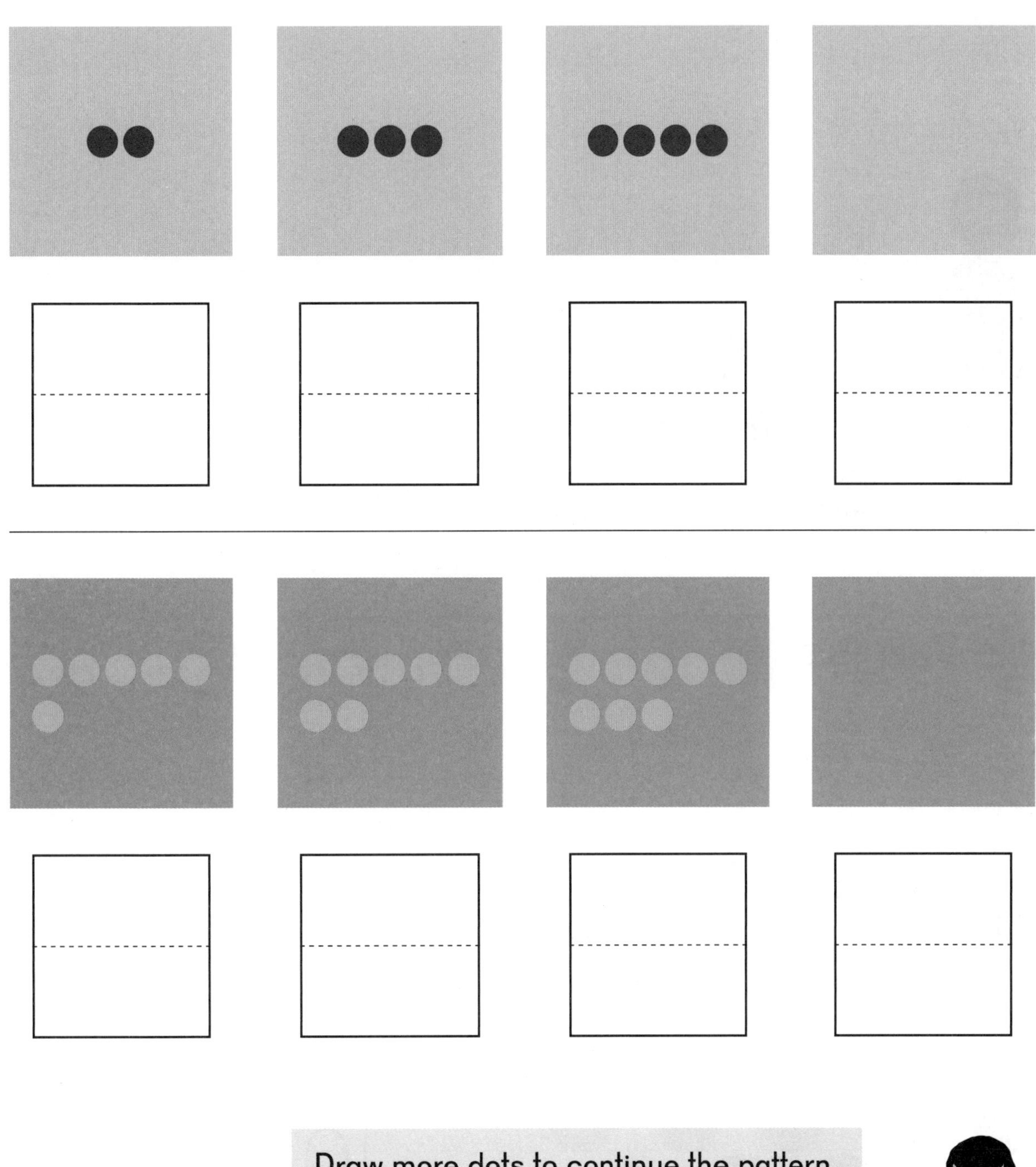

Draw more dots to continue the pattern.
Write the numbers.

Objective: Recognize that numbers in order increase by 1.

Exercise 11 • page 71

Lesson 13
Practice

P 13

Circle how many.

4 6 10

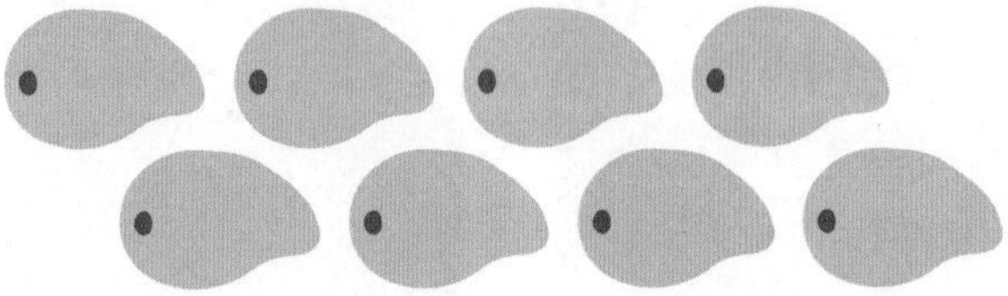

10 9 8

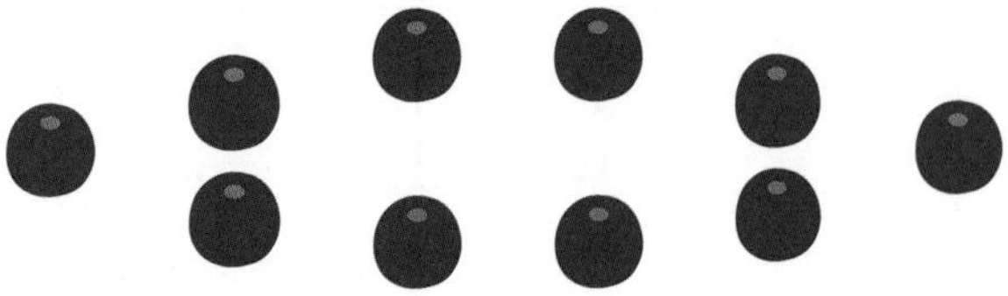

5 10 7

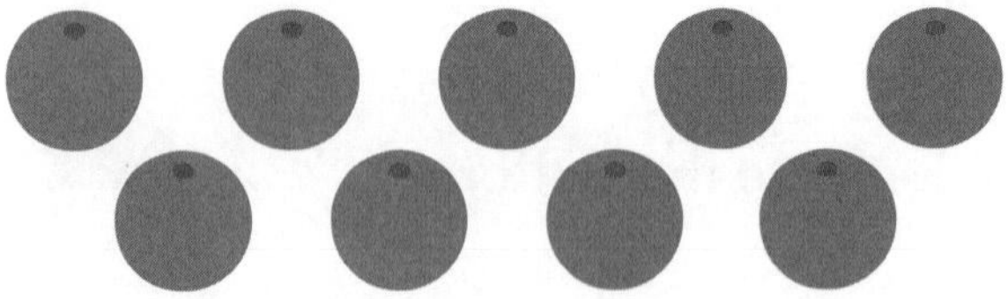

10 9 8

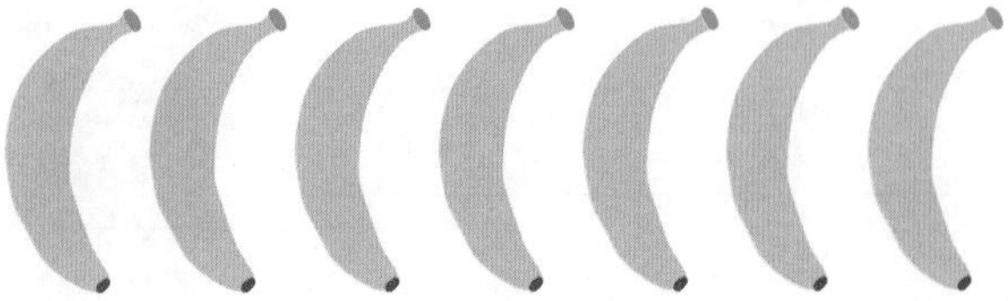

7 8 9

Objective: Practice.

Count and write how many.

Objective: Practice.

Draw 1 more and write the number.

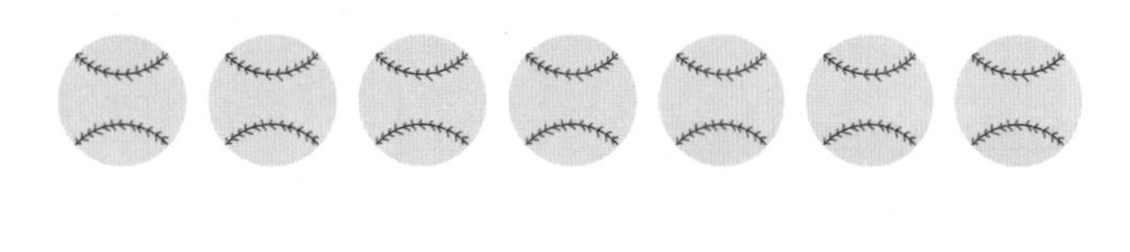

Objective: Practice.

Exercise 12 • page 75

Chapter 4

Shapes and Solids

Lesson 1
Curved or Flat

1

Look and talk.

Are the surfaces curved or flat?

Objective: Identify curved and flat surfaces and build with some solids.

Lesson 2
Solid Shapes

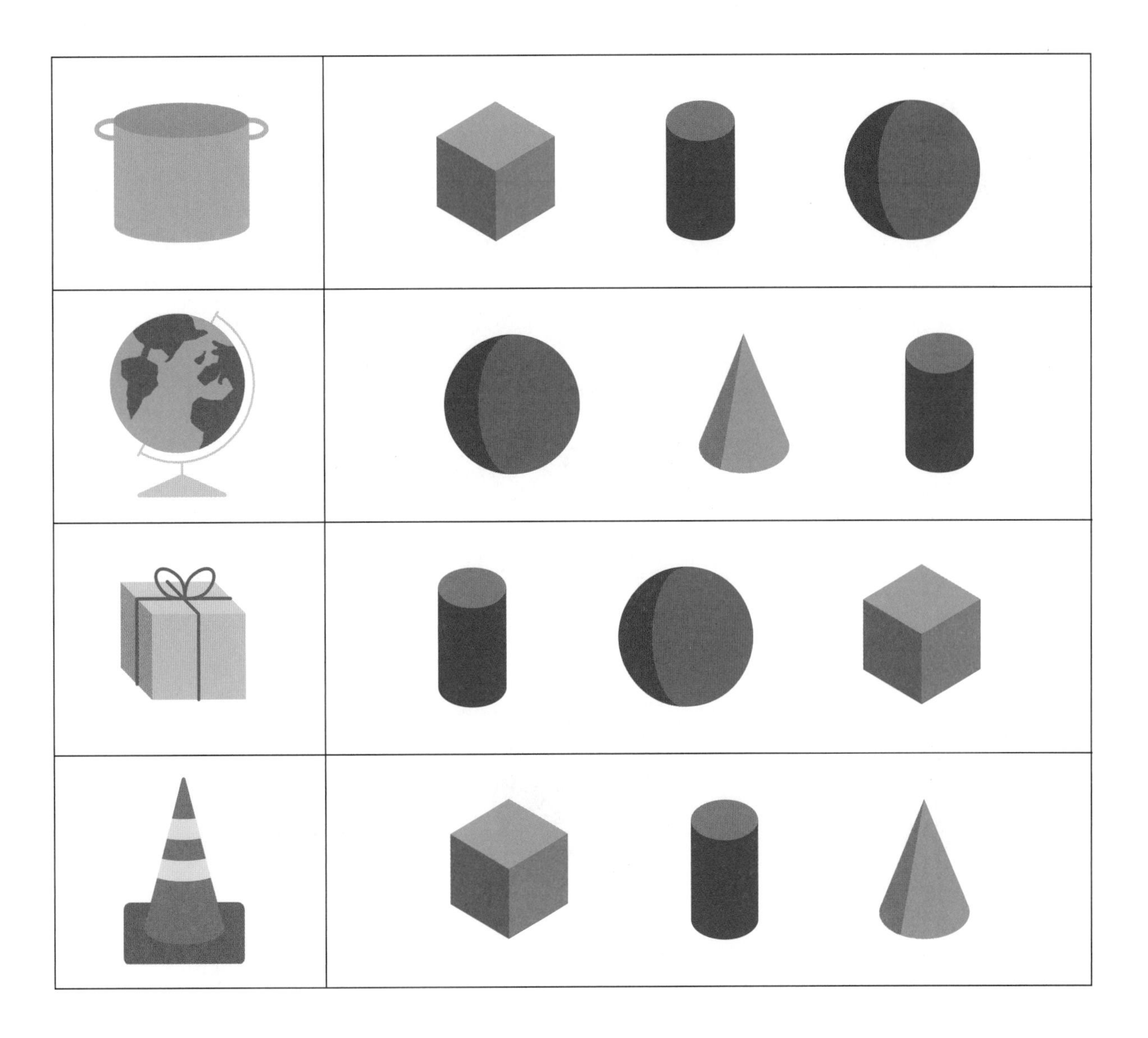

Circle the solid that is similar to the first one in each row.

Objective: Recognize cubes, cylinders, spheres, and cones.

Cross out the one that does not belong in each row.

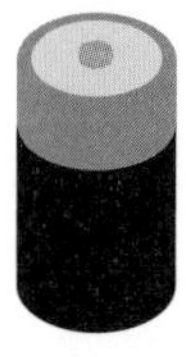

Objective: Recognize cubes, cylinders, spheres, and cones.

Exercise 1 • page 81

Look and talk.

Which shapes are closed?

Objective: Identify closed shapes.

Exercise 2 • page 83

Lesson 4
Rectangles

Objective: Recognize rectangles.

Objective: Recognize rectangles.

Exercise 3 • page 85

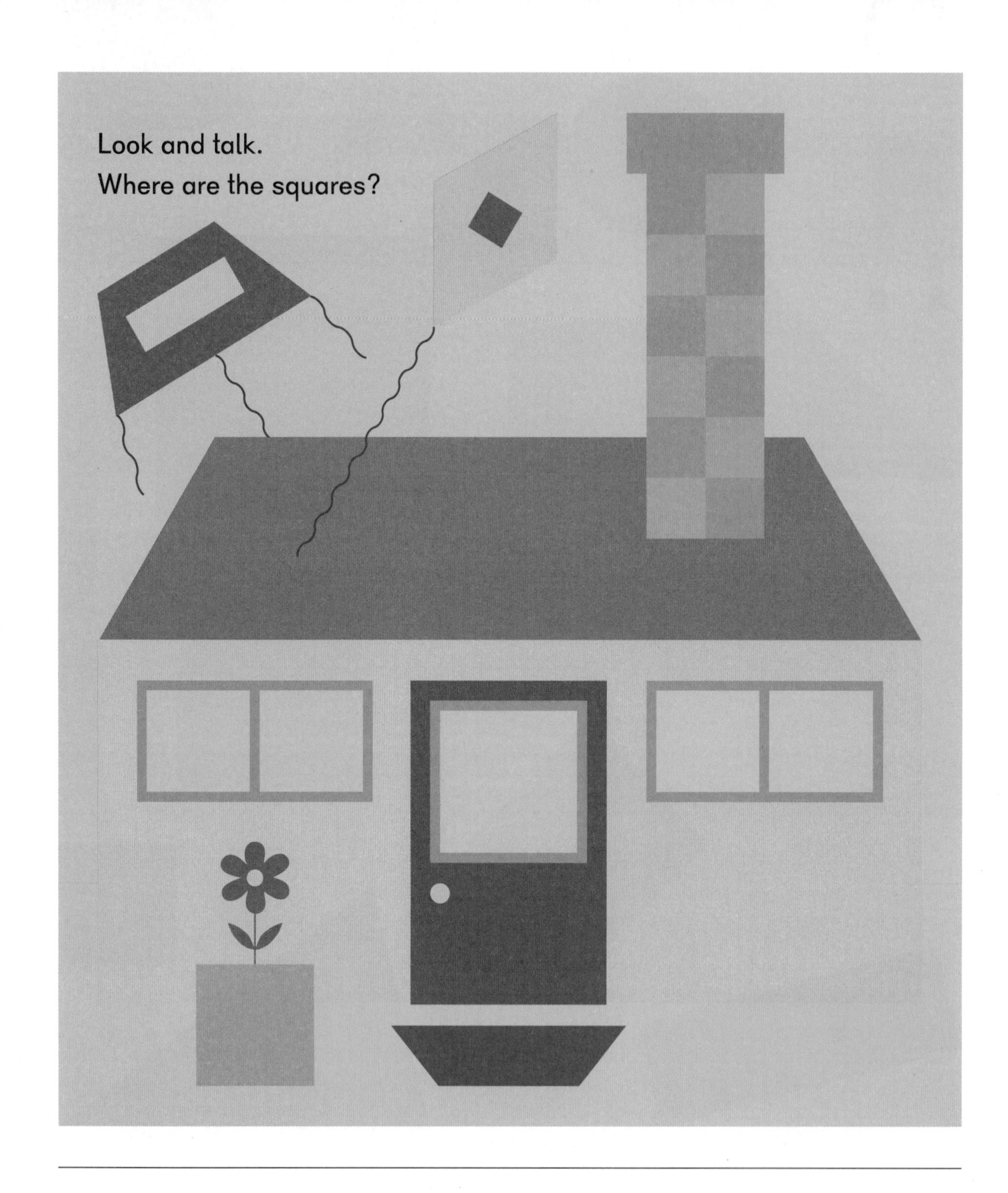

Objective: Identify squares.

Exercise 4 • page 87

Objective: Identify circles and triangles.

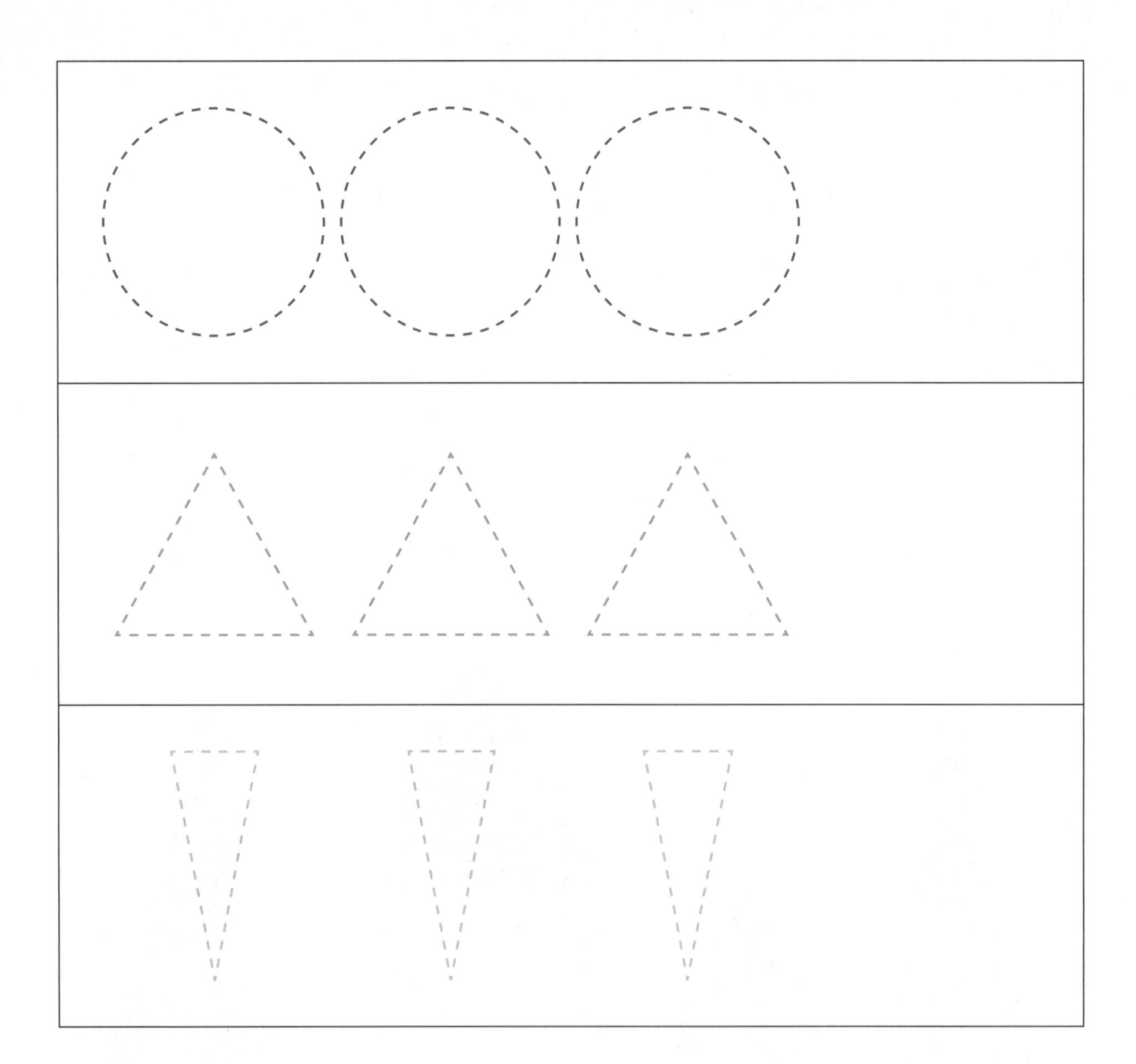

Trace each shape and draw one more.

Objective: Identify, name, and draw circles and triangles.

Objective: Recognize triangles as being closed figures with 3 straight sides and 3 corners.

Exercise 5 • page 89

Lesson 7
Where is It?

7

Objective: Use position words to describe the location of an object.

Draw a ▲ above Alex.
Draw a ■ below Alex.
Draw a ▲ near the ■.

Objective: Follow directions and draw shapes in given positions.

Exercise 6 • page 93

Copy this picture with your pattern blocks.

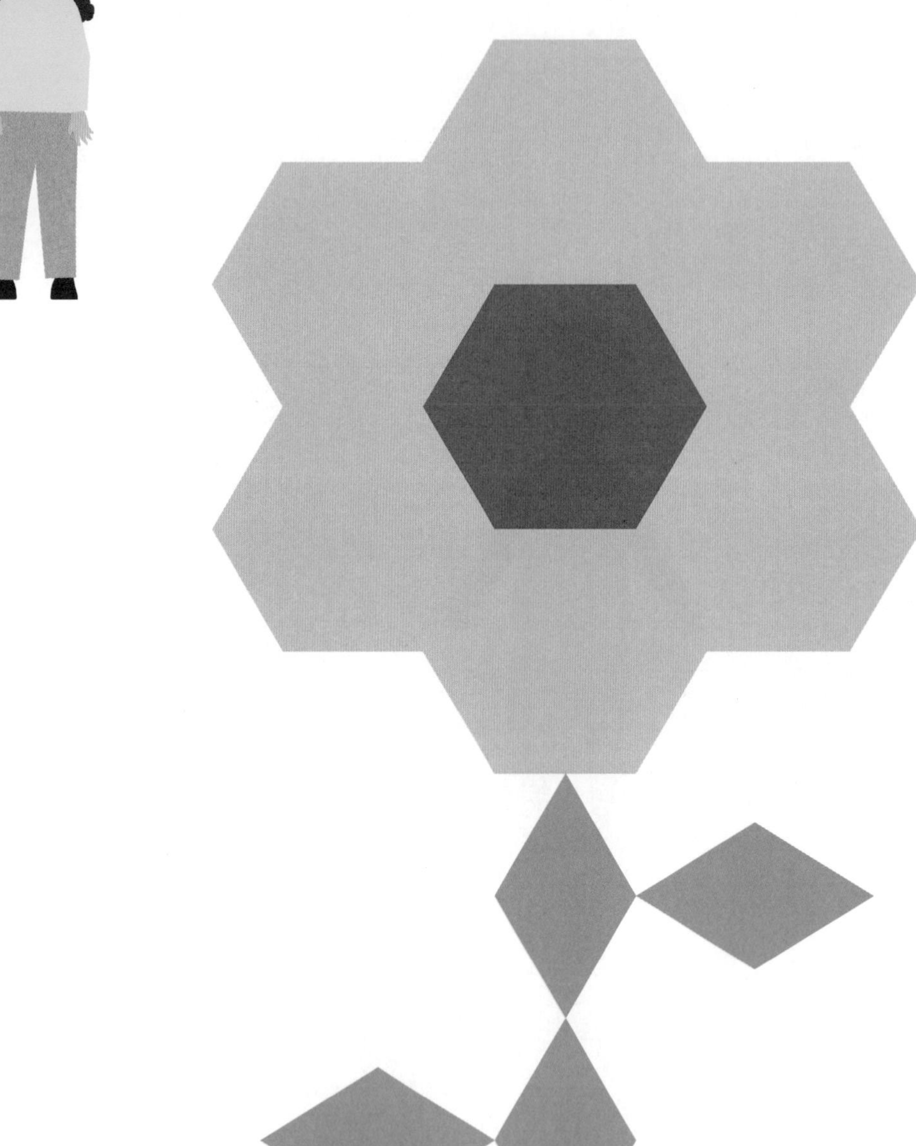

Objective: Recognize and name hexagons and copy a pattern.

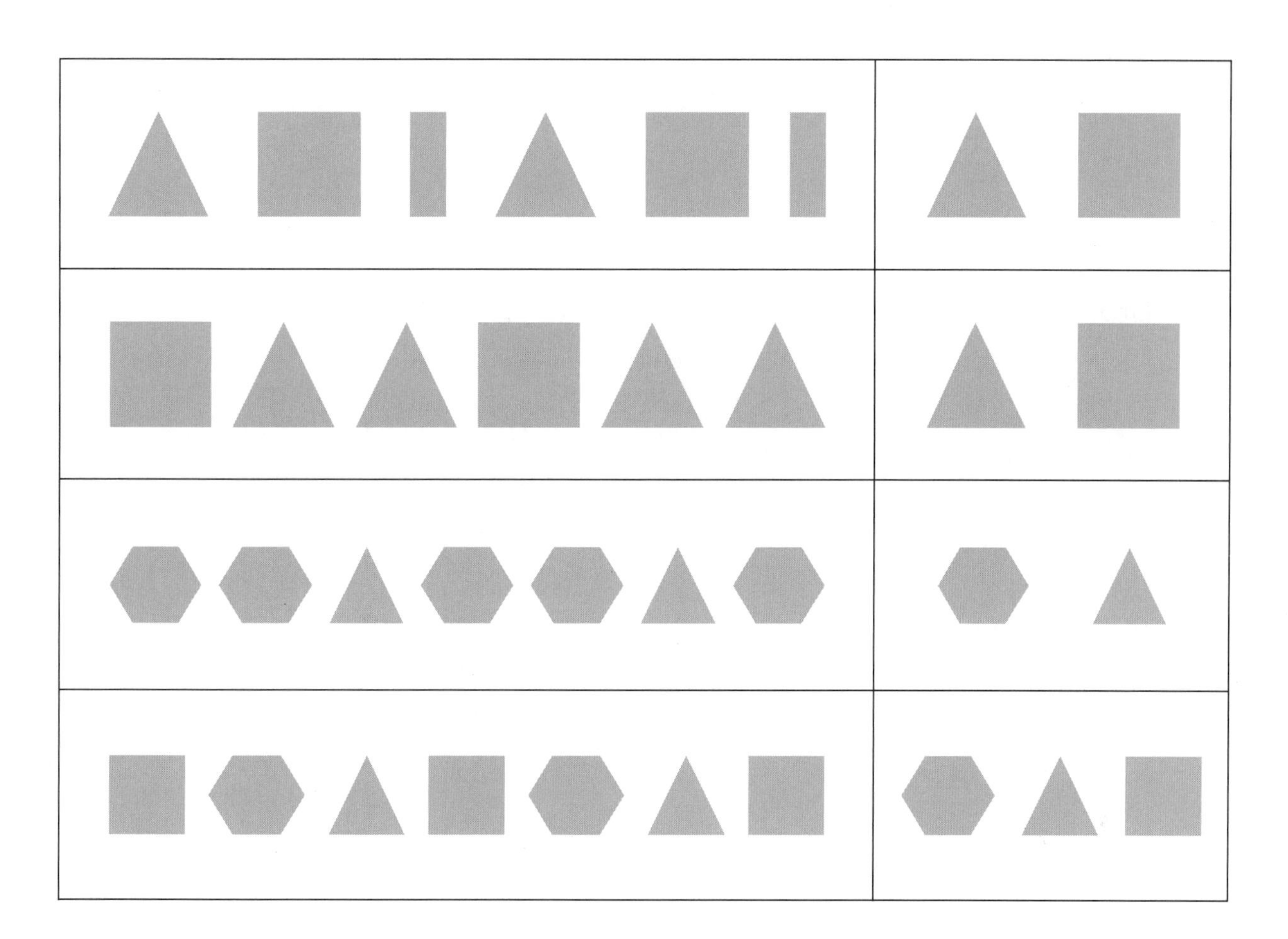

Say the name of the shapes,
then circle the shape that comes next.

Objective: Continue patterns.

Exercise 7 • page 95

Lesson 9
Sizes and Shapes

9

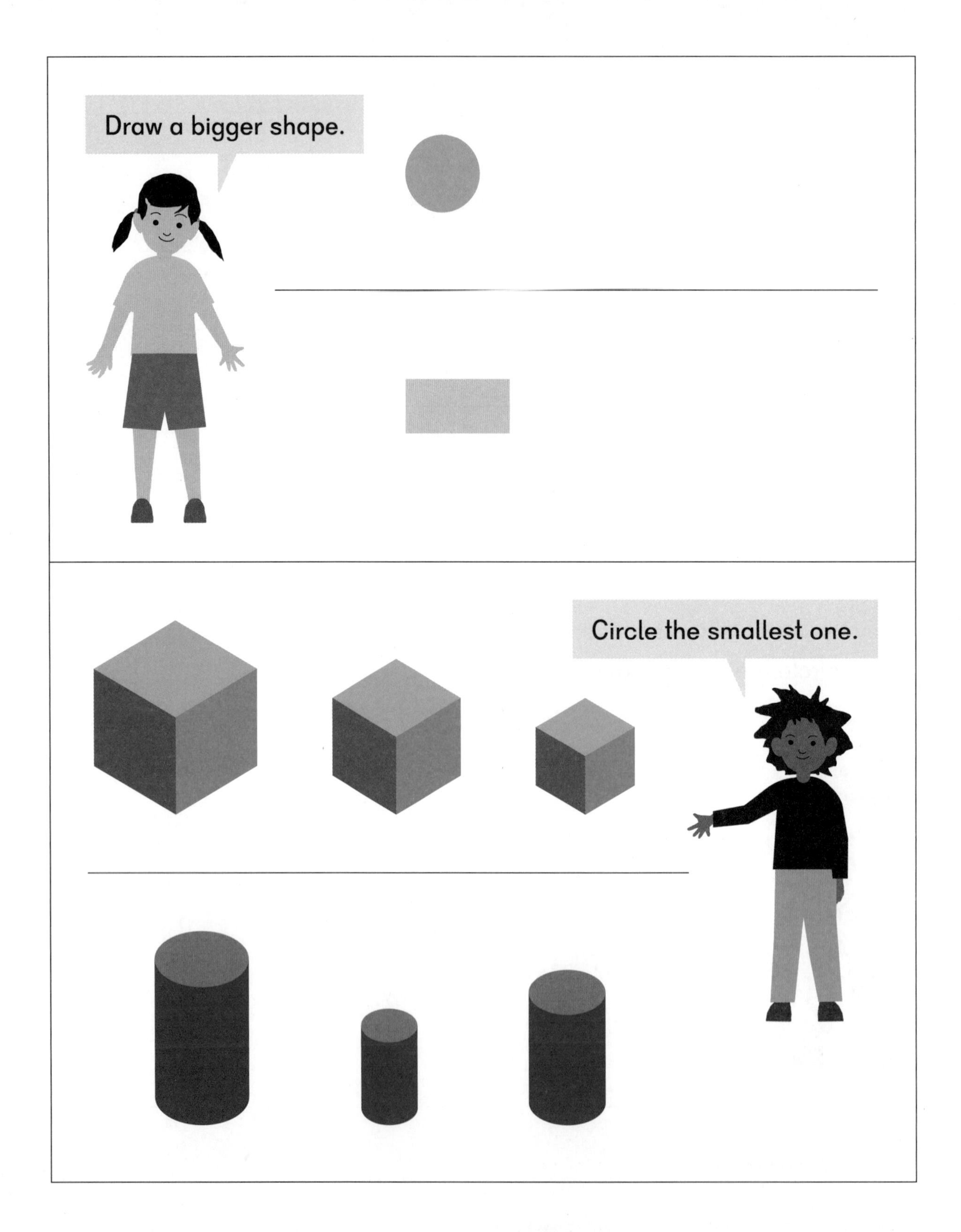

Objective: Identify and draw shapes according to size and identify solids by size.

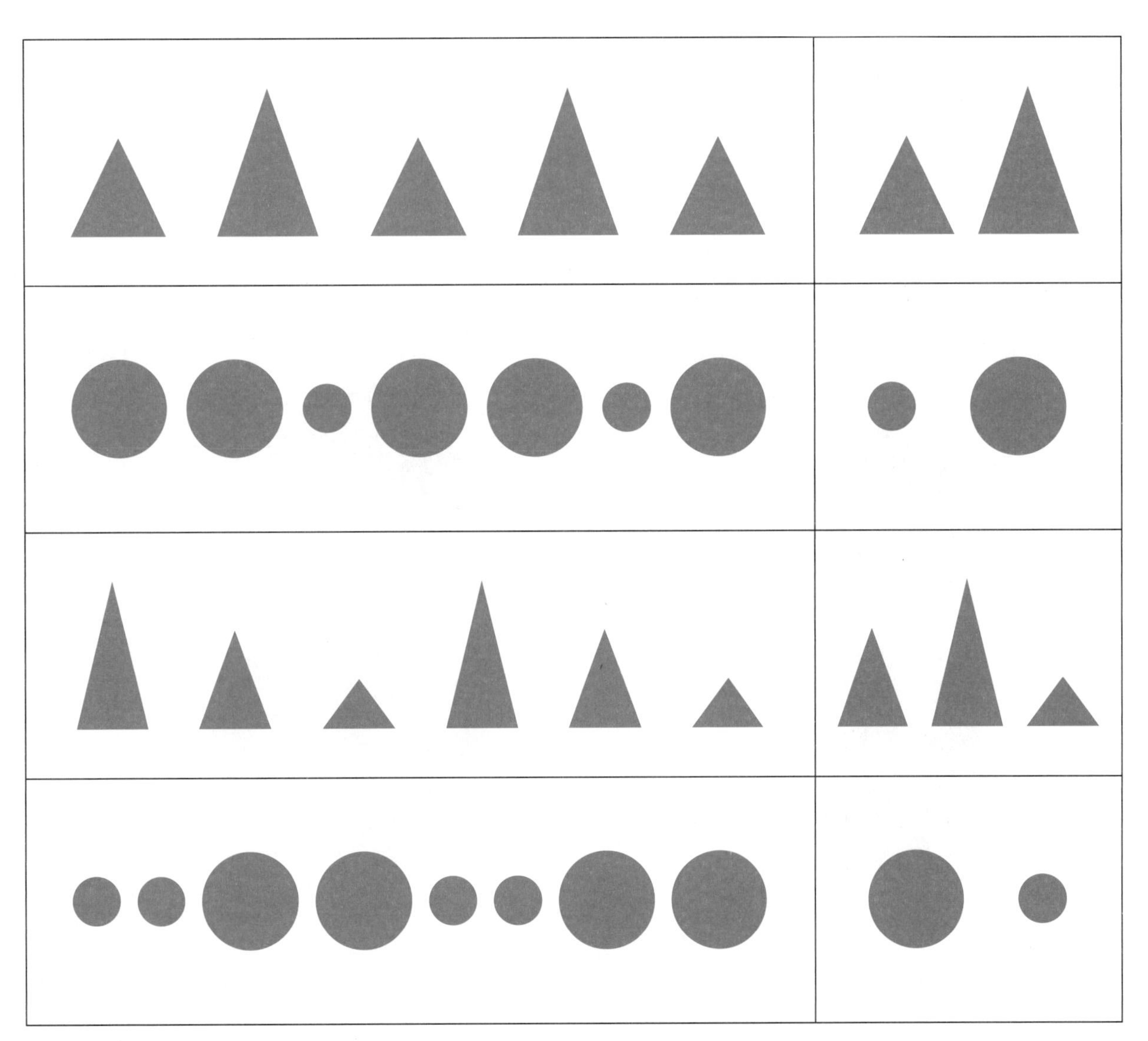

Circle the one that comes next.

Objective: Continue patterns.

Exercise 8 • page 97

Look and talk.

What shapes are Alex and Emma using to make other shapes?

Objective: Make new shapes by combining basic shapes.

Objective: Combine simple shapes to make larger shapes.

Exercise 9 • page 99

Lesson 11
Graphs

11

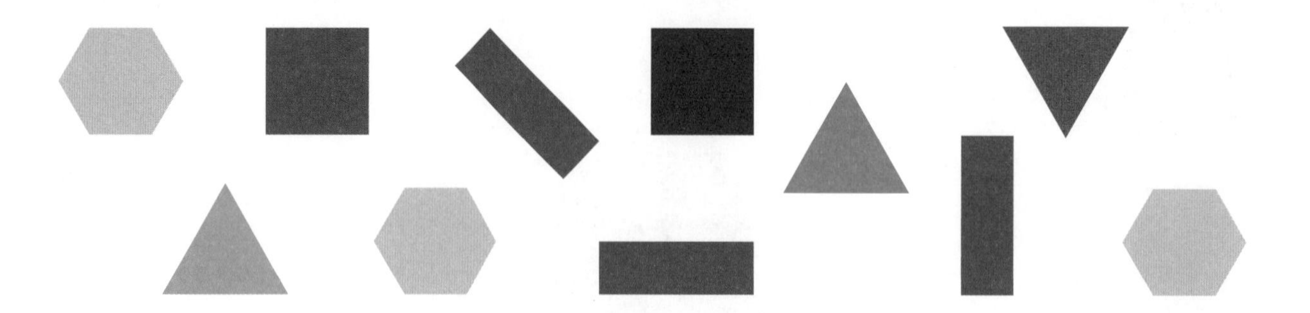

Count the shapes.

Color the graph to show how many.

Write how many.

Shapes			

Objective: Represent data in a graph.

Count the solids.

Color the graph to show how many.

Write how many.

Solids			

Objective: Represent data in a graph.

Exercise 10 • page 101

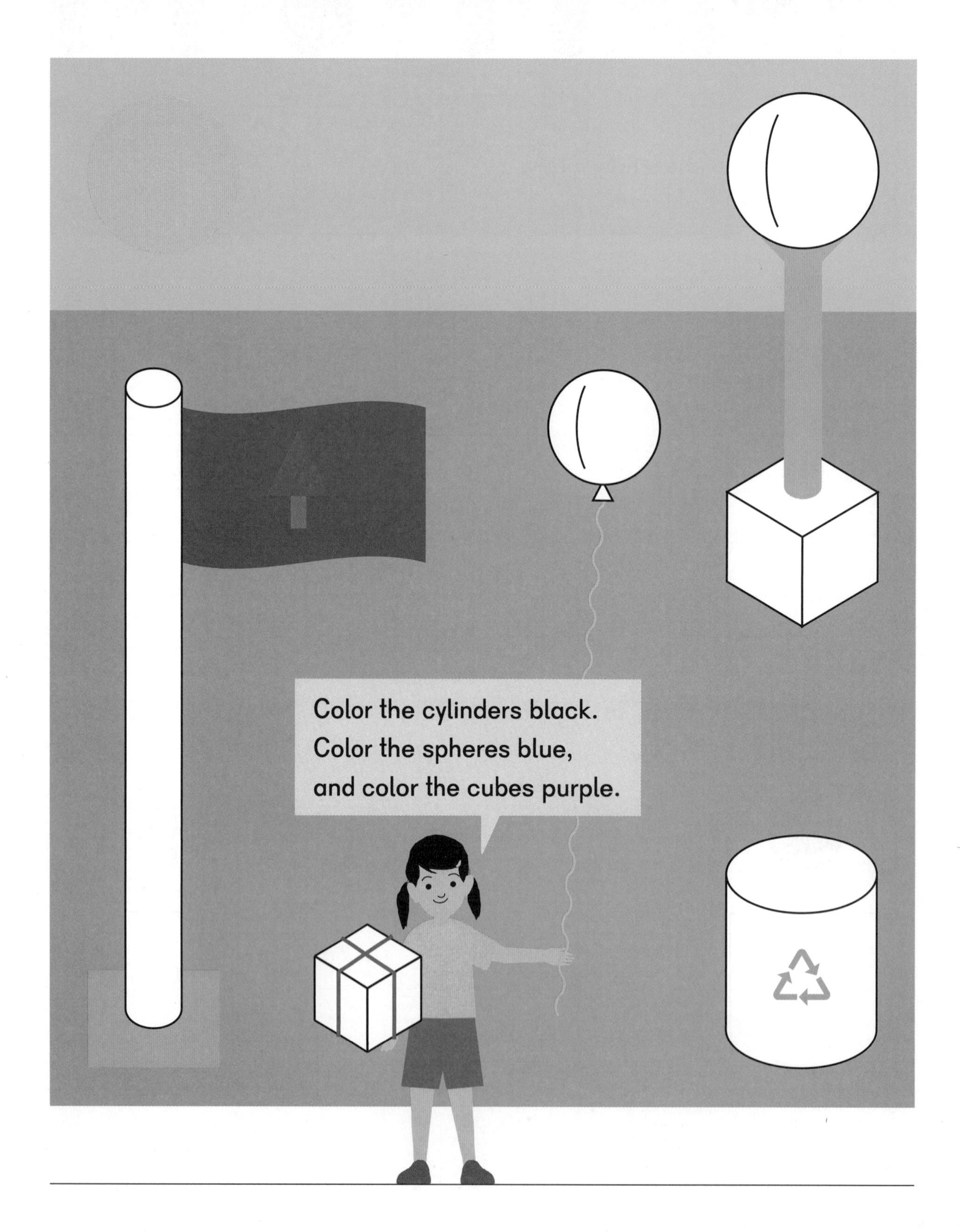

Objective: Practice.

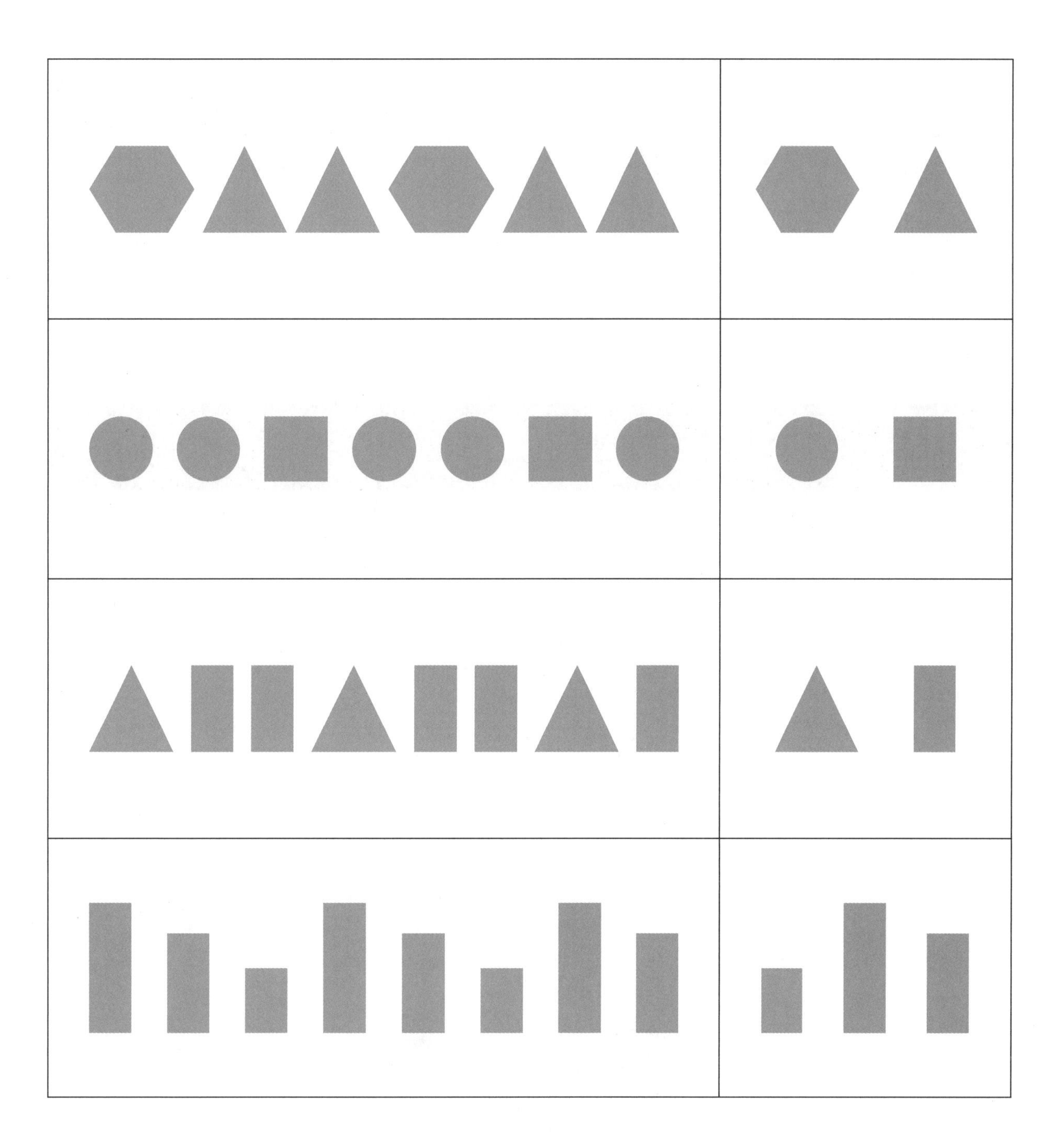

Circle the shape that comes next.

Objective: Practice.

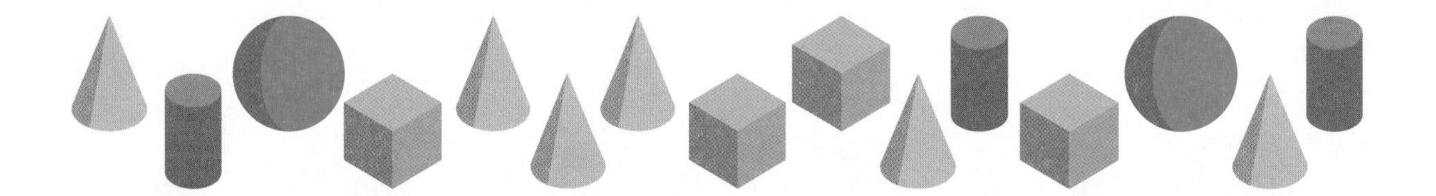

Count the solids.

Color the graph to show how many.

Write how many.

Solids			

Objective: Practice.

Exercise 11 • page 103

Chapter 5

Compare Height, Length, Weight, and Capacity

Lesson 1
Comparing Height

1

Look and talk.
Who is taller?

Objective: Compare height.

Objective: Compare height.

Cross out the tallest tree in each row.
Circle the shortest tree in each row.

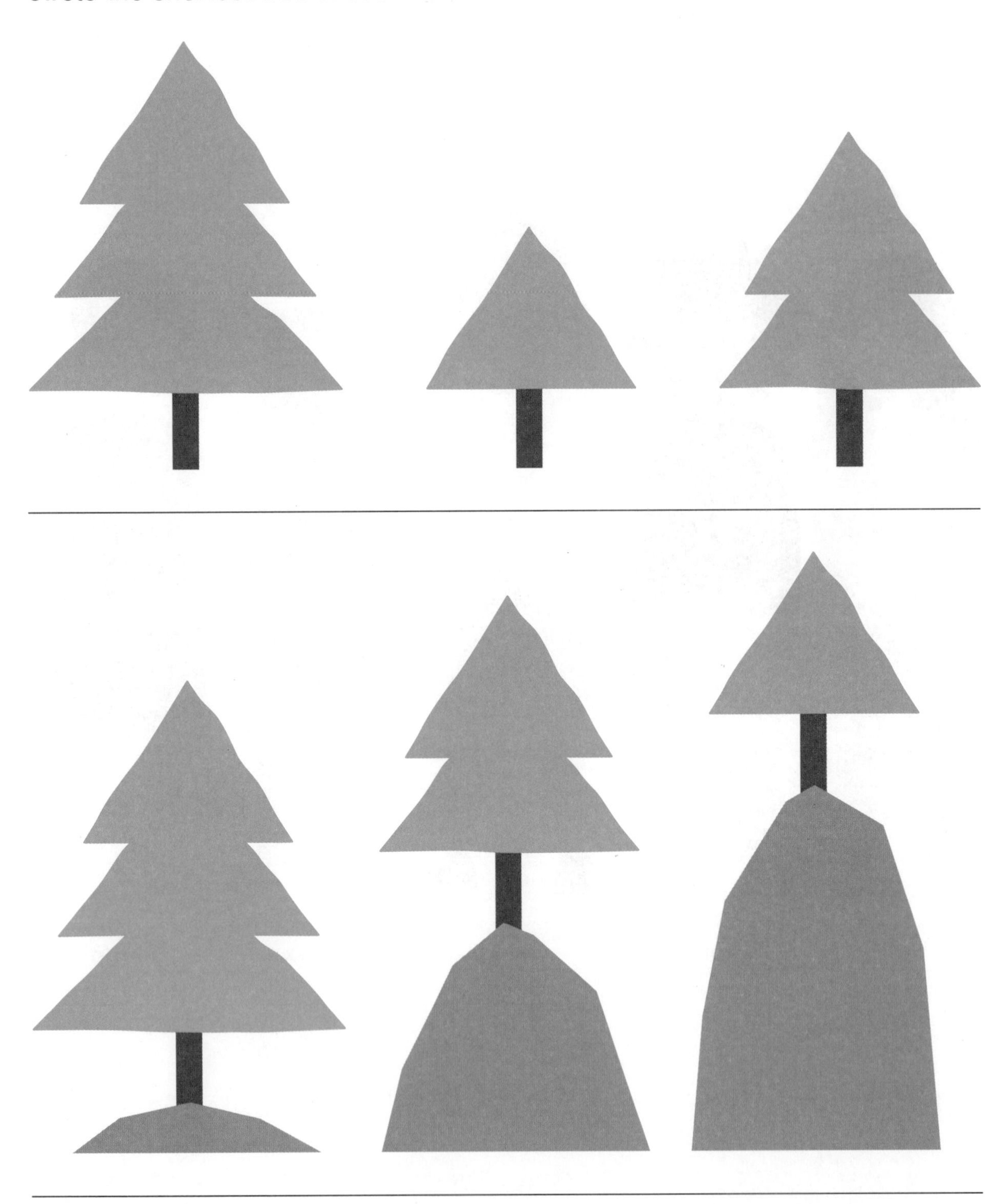

Objective: Compare height.

Exercise 1 • page 107

Lesson 2
Comparing Length

2

Objective: Compare length.

Draw a line that is longer than mine and a line that is shorter than mine.

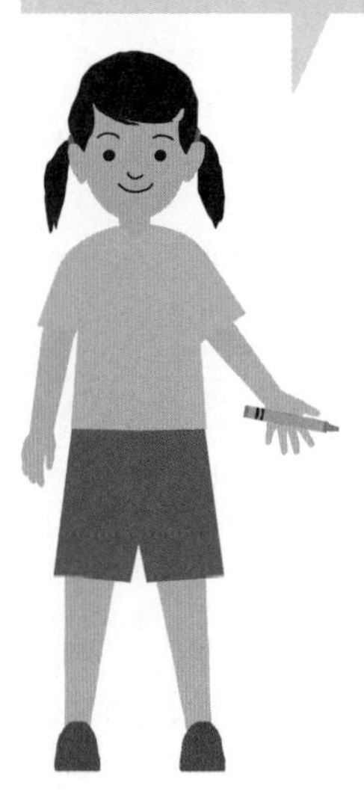

Longer

↓

↑

Shorter

Objective: Compare length.

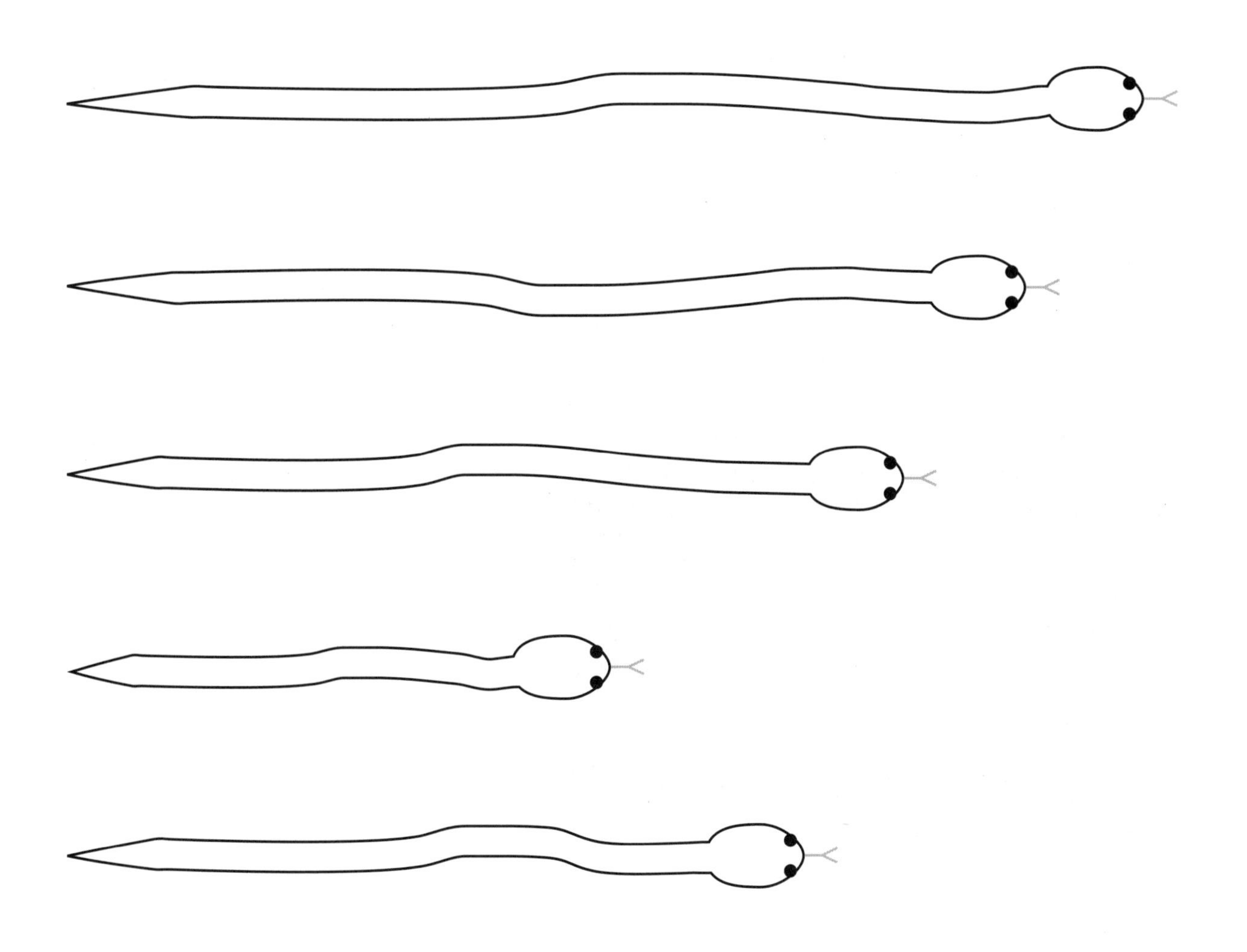

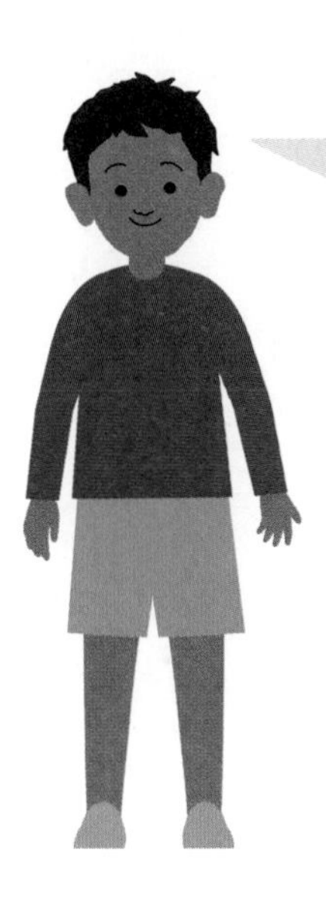

Color the longest snake blue.
Color the shortest snake green.

Objective: Compare length.

Exercise 2 • page 109

Lesson 3
Height and Length — Part 1

3

Objective: Compare length of objects indirectly using a third object.

Objective: Measure objects using non-standard units.

Exercise 3 • page 111

Lesson 4
Height and Length — Part 2

Use your linking cubes to measure my hair ribbons.

Objective: Measure objects using non-standard units.

How tall is each bottle?

Count the cubes and color them in for each bottle.

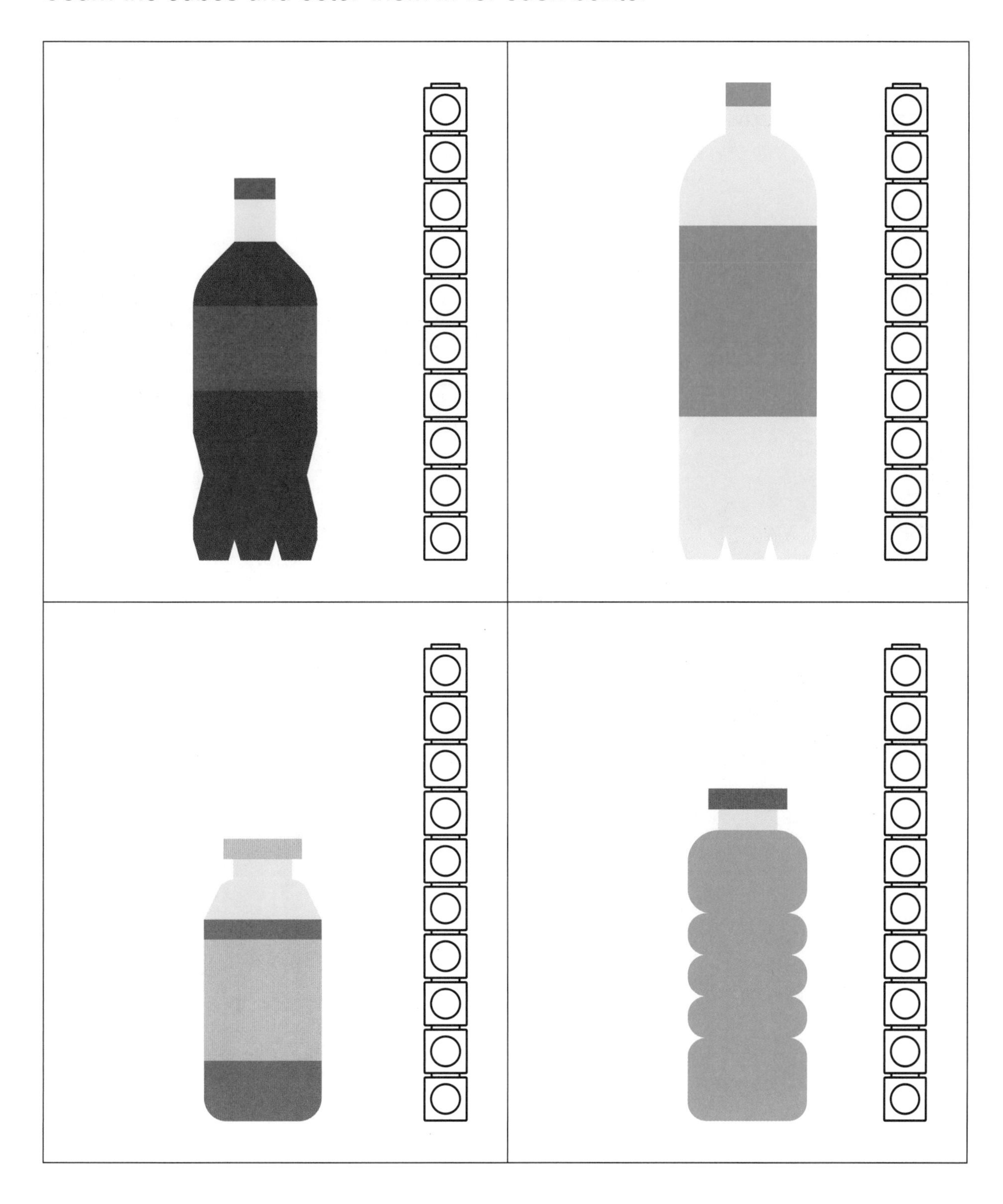

Objective: Measure objects using non-standard units.

Exercise 4 • page 113

Lesson 5
Weight — Part 1

5

Look and talk.
Which thing is heavier?

Objective: Compare weight.

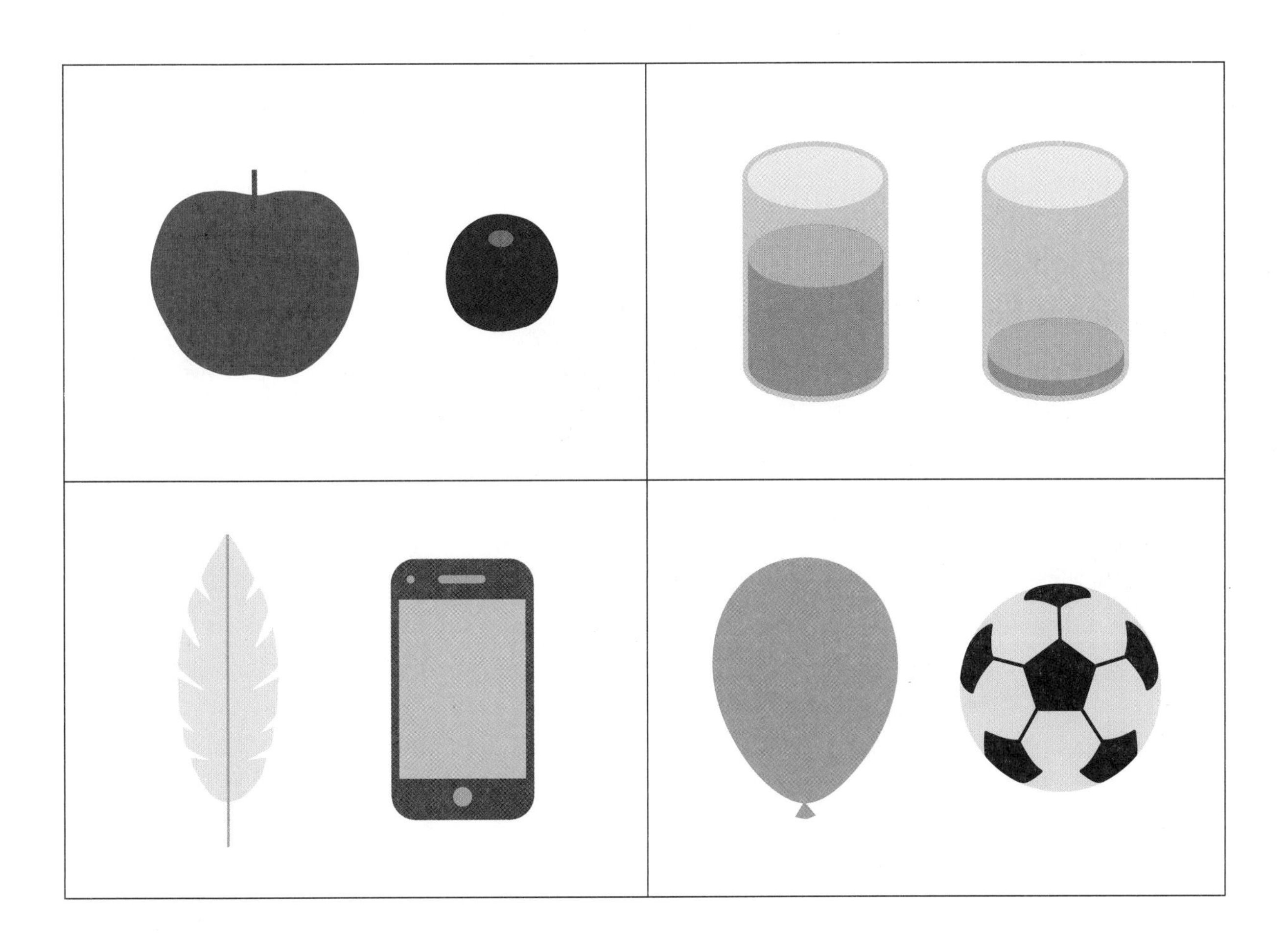

This book is heavier than my juice box.
Circle the thing that is heavier.

Objective: Compare weight.

Exercise 5 • page 117

Lesson 6
Weight — Part 2

6

Look and talk.
Which one is heavier?
Which one is lighter?

Objective: Compare weight.

Exercise 6 • page 119

This pencil weighs the same as 2 linking cubes.
Look at the pictures and write the numbers in the boxes.

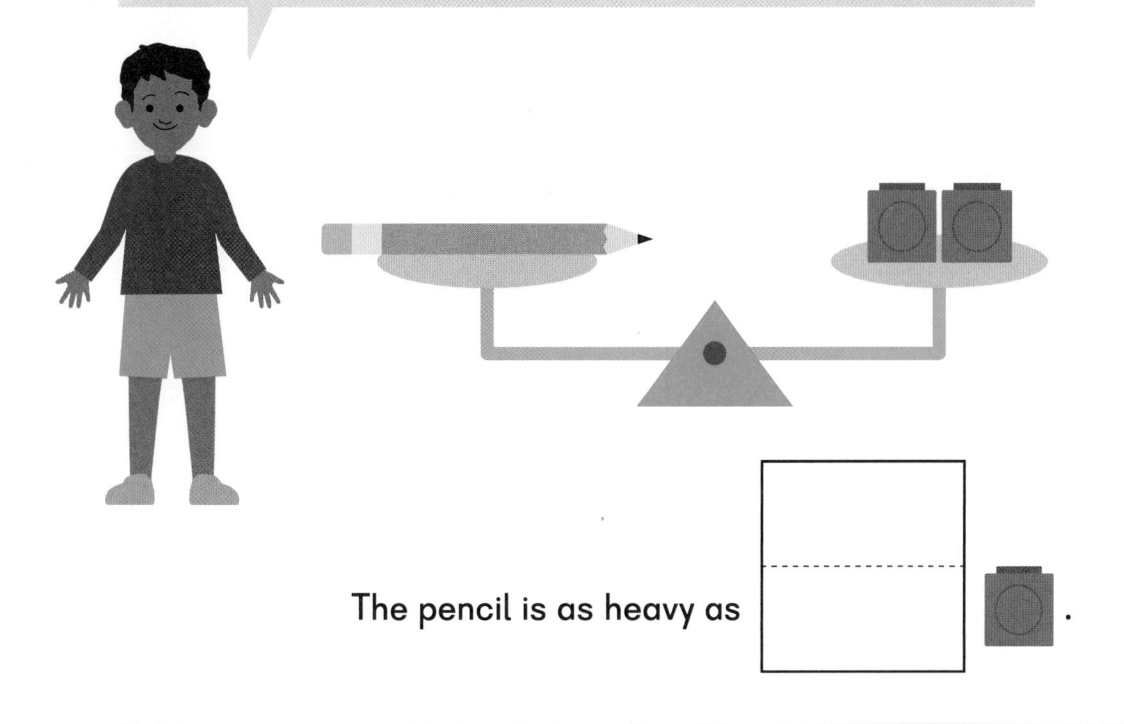

The pencil is as heavy as ☐ cubes.

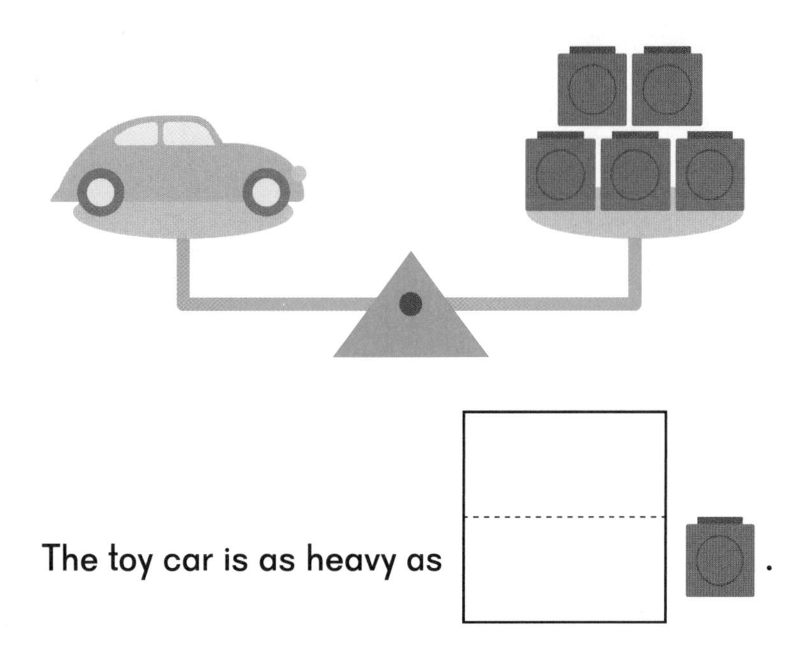

The toy car is as heavy as ☐ cubes.

Objective: Compare weight and measure weight with non-standard units.

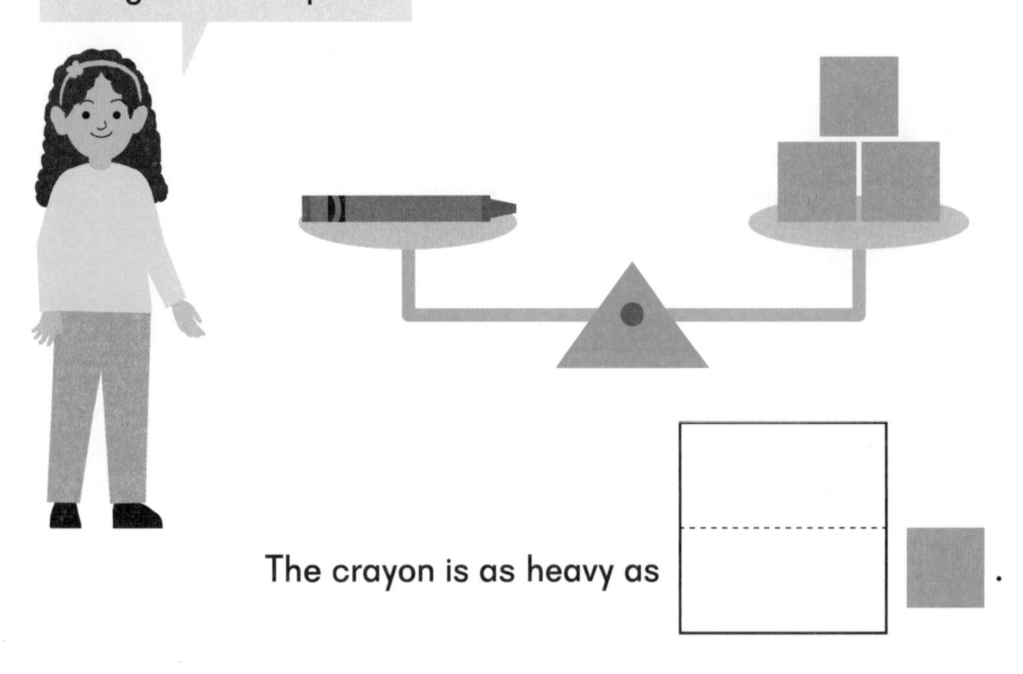

The crayon is as heavy as ______ .

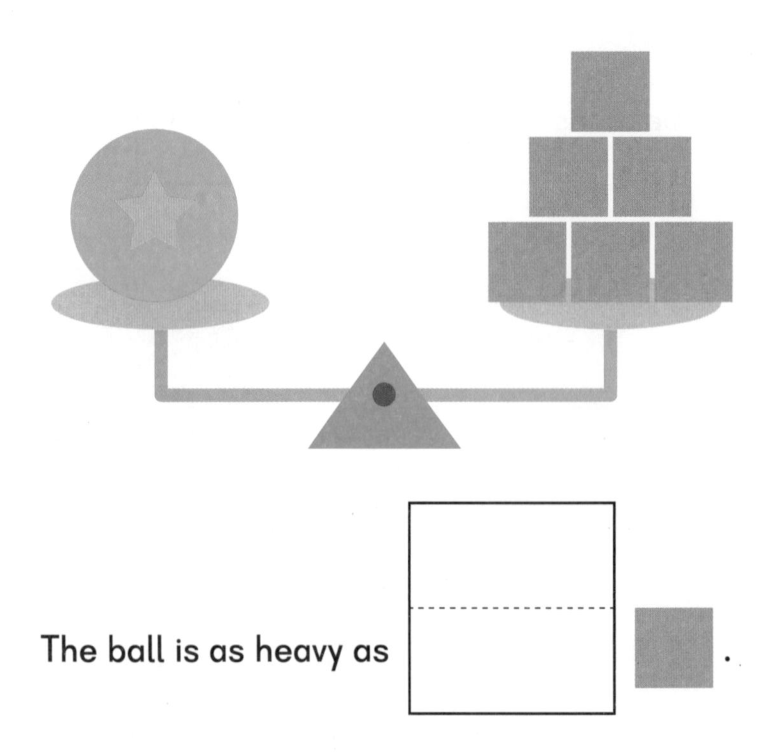

The ball is as heavy as ______ .

Objective: Compare weight and measure weight with non-standard units.

Exercise 7 • page 121

Lesson 8
Capacity — Part 1

8

Look and talk.
Which holds more?

Objective: Compare capacity.

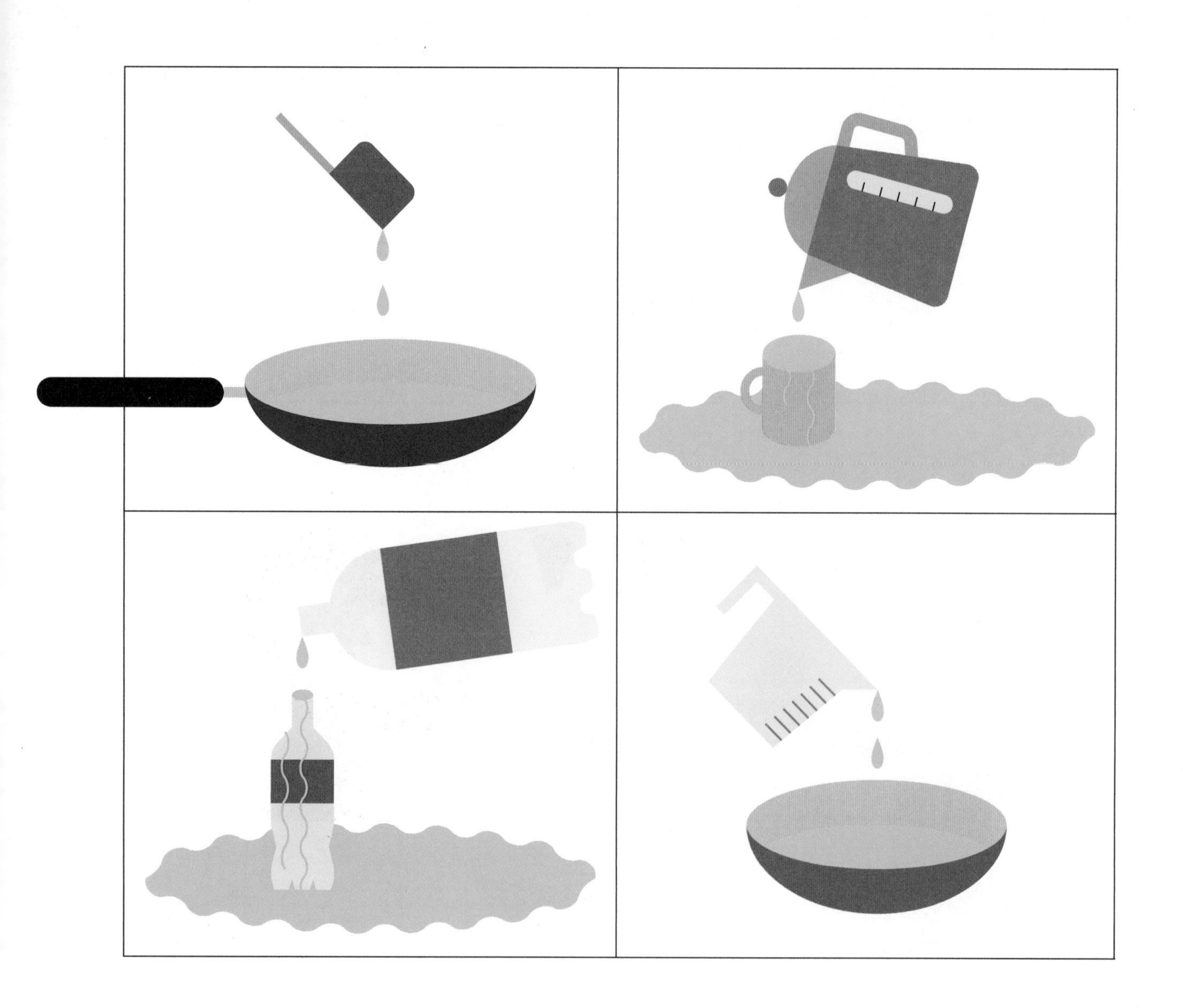

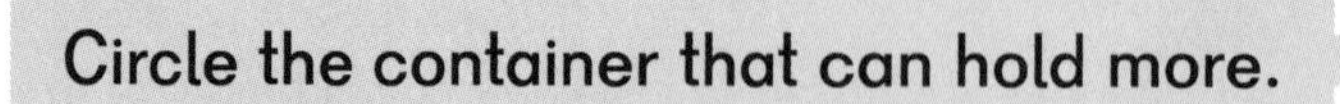

Circle the container that can hold more.

Objective: Compare capacity.

Exercise 8 • page 123

Lesson 9
Capacity — Part 2

Objective: Measure capacity with non-standard units.

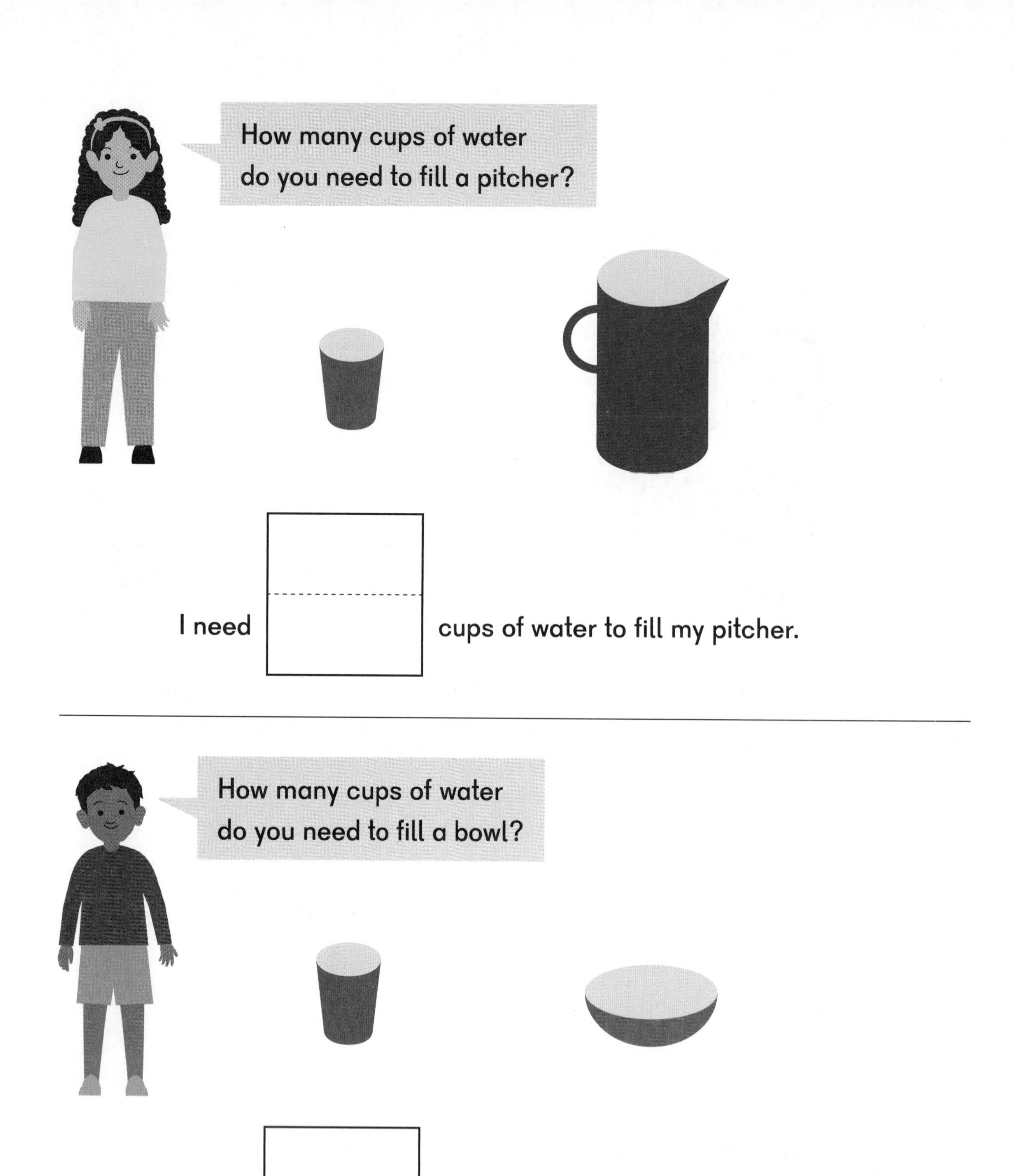

Objective: Measure capacity with non-standard units.

Exercise 9 • page 125

Which set of pencils is shown in order from shortest to longest?

Objective: Practice.

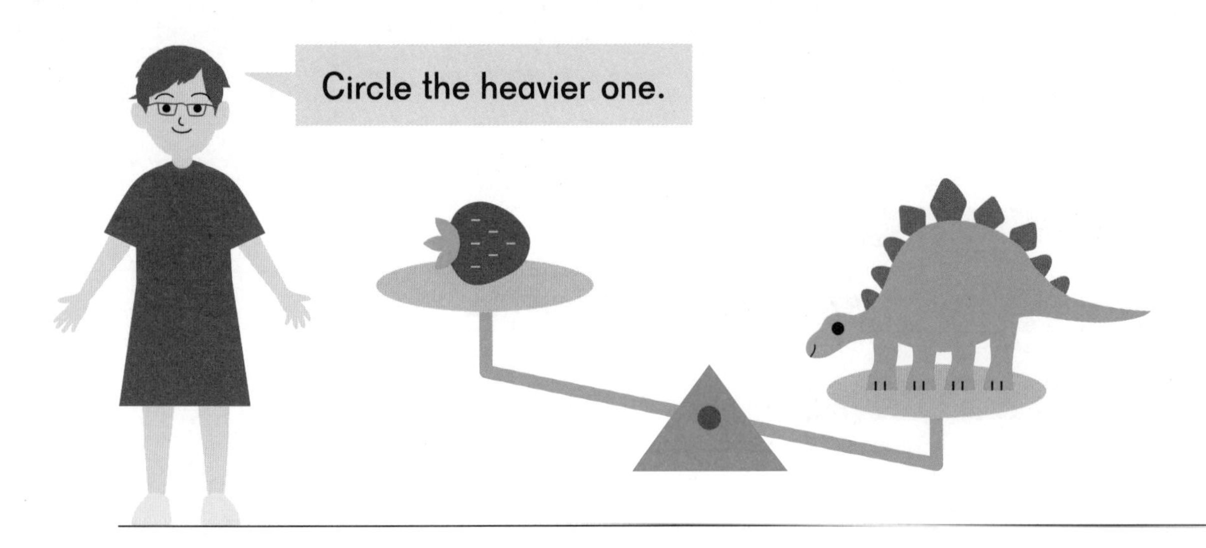

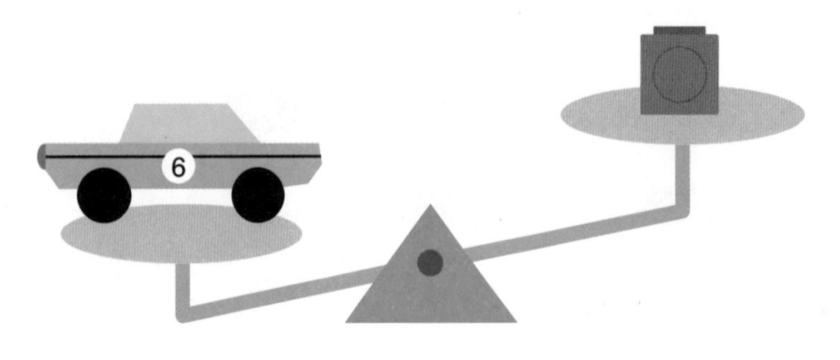

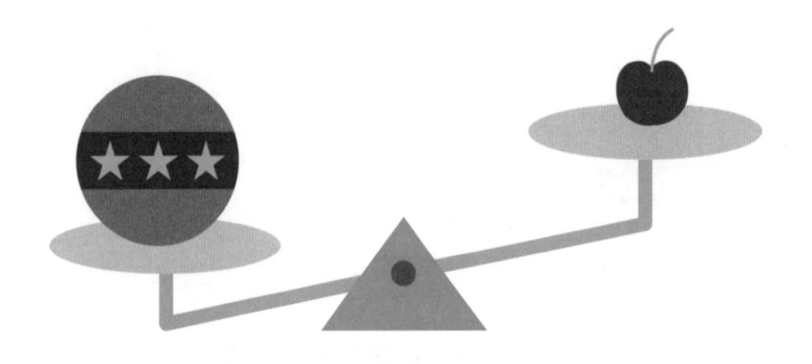

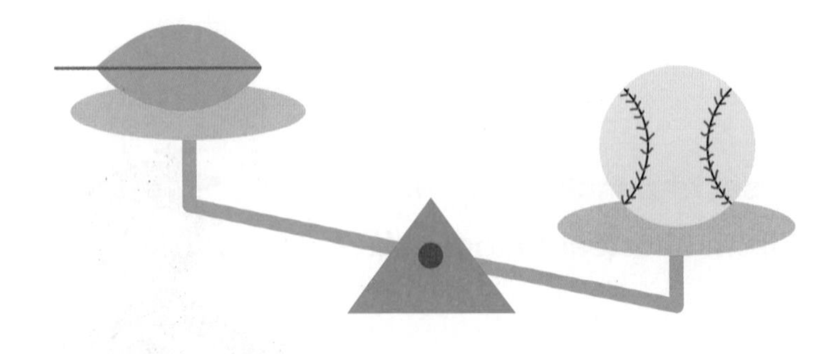

Objective: Practice.

Exercise 10 • page 127

Chapter 6

Comparing Numbers Within 10

Objective: Recognize equal groups.

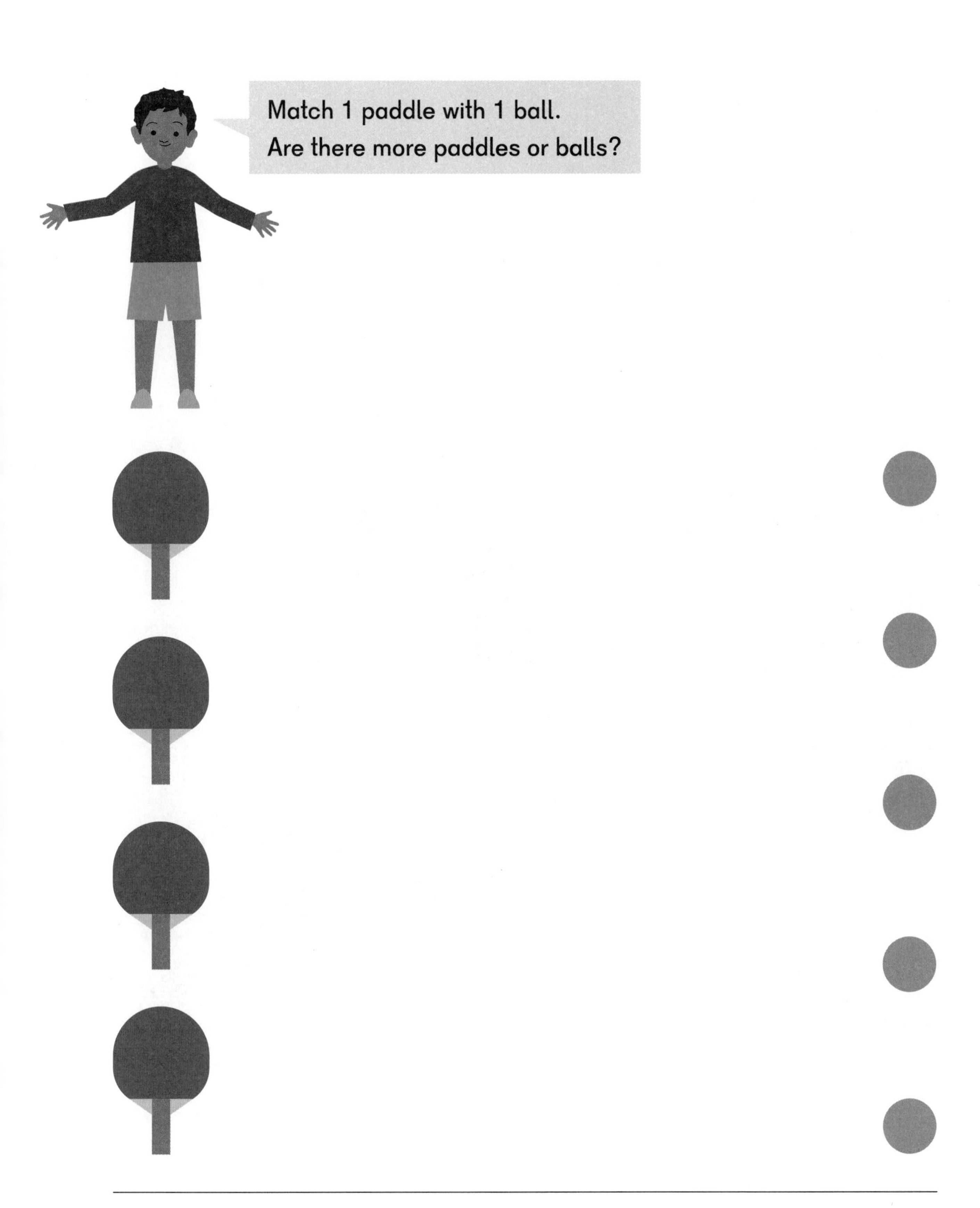

Objective: Identify a group that has more objects than another group.

Exercise 1 • page 129

Lesson 2
More and Fewer

Objective: Identify a group which has more and a group which has fewer objects than another group.

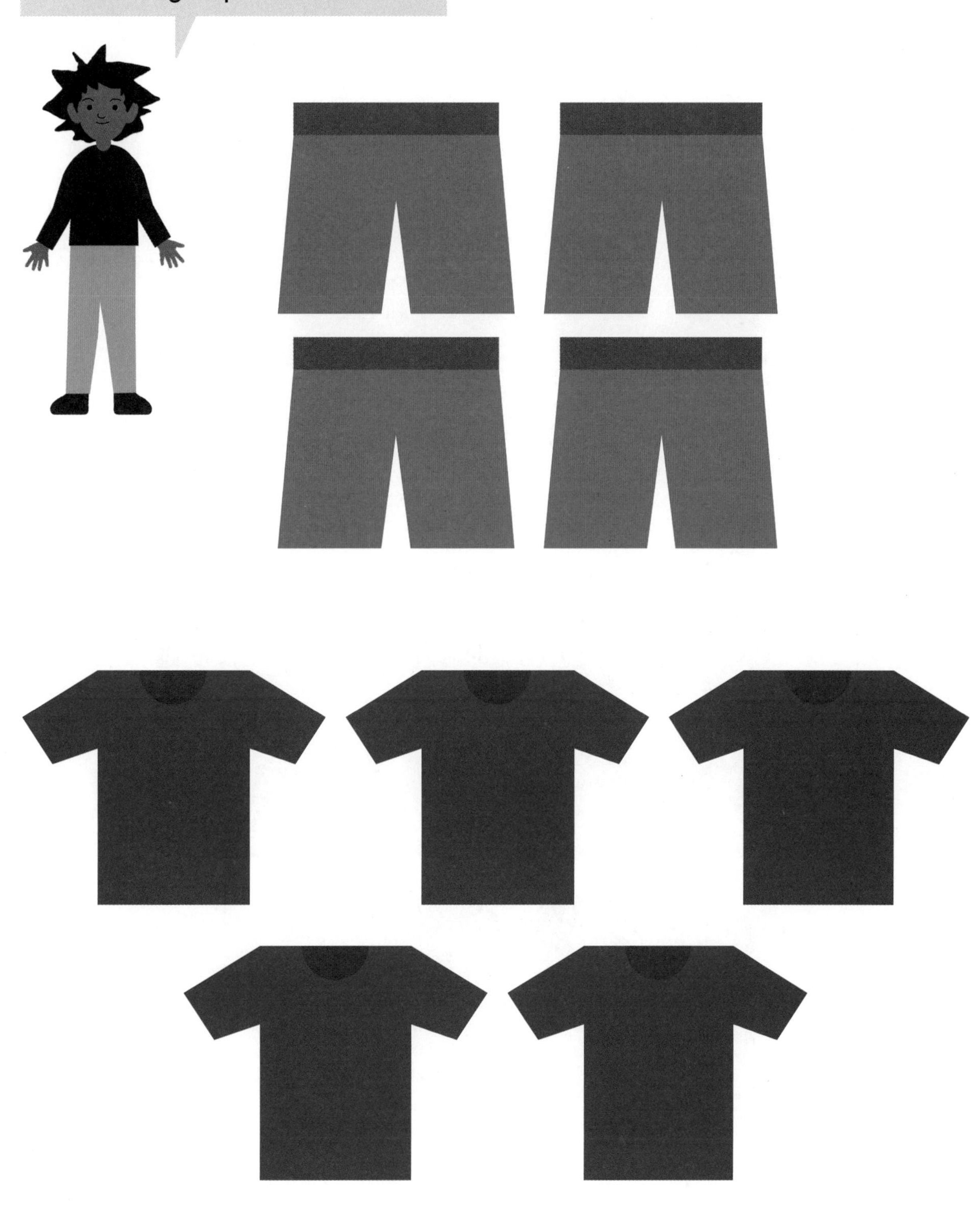

Objective: Identify a group which has more than another group.

Objective: Identify a group that has fewer objects than another group.

Exercise 2 • page 133

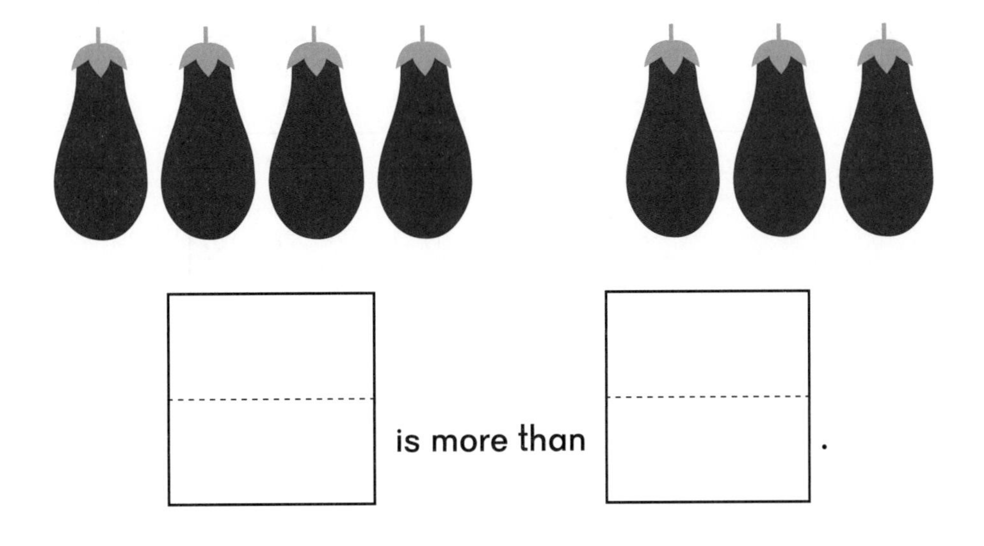

is more than .

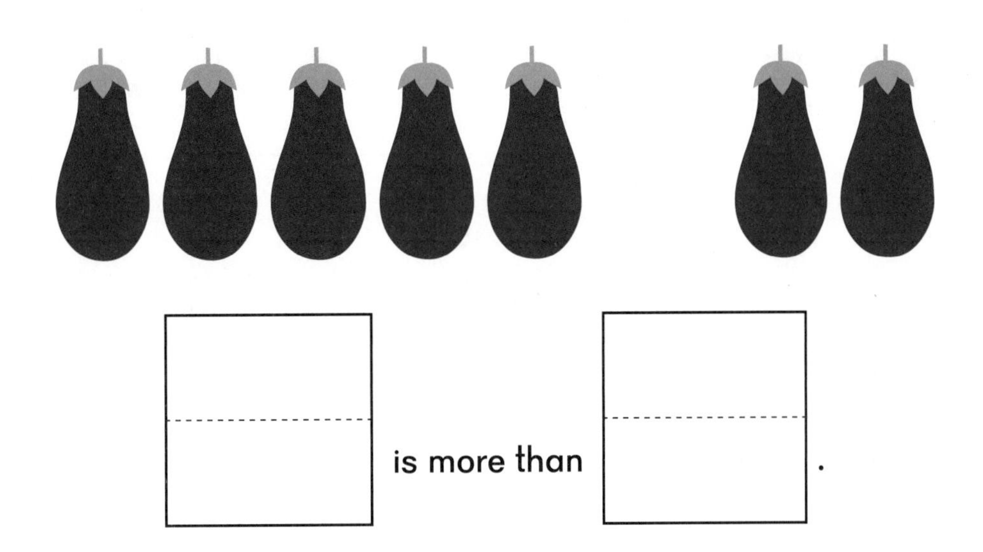

is more than .

Write the numbers.

Objective: Compare sets of objects and numbers using "more."

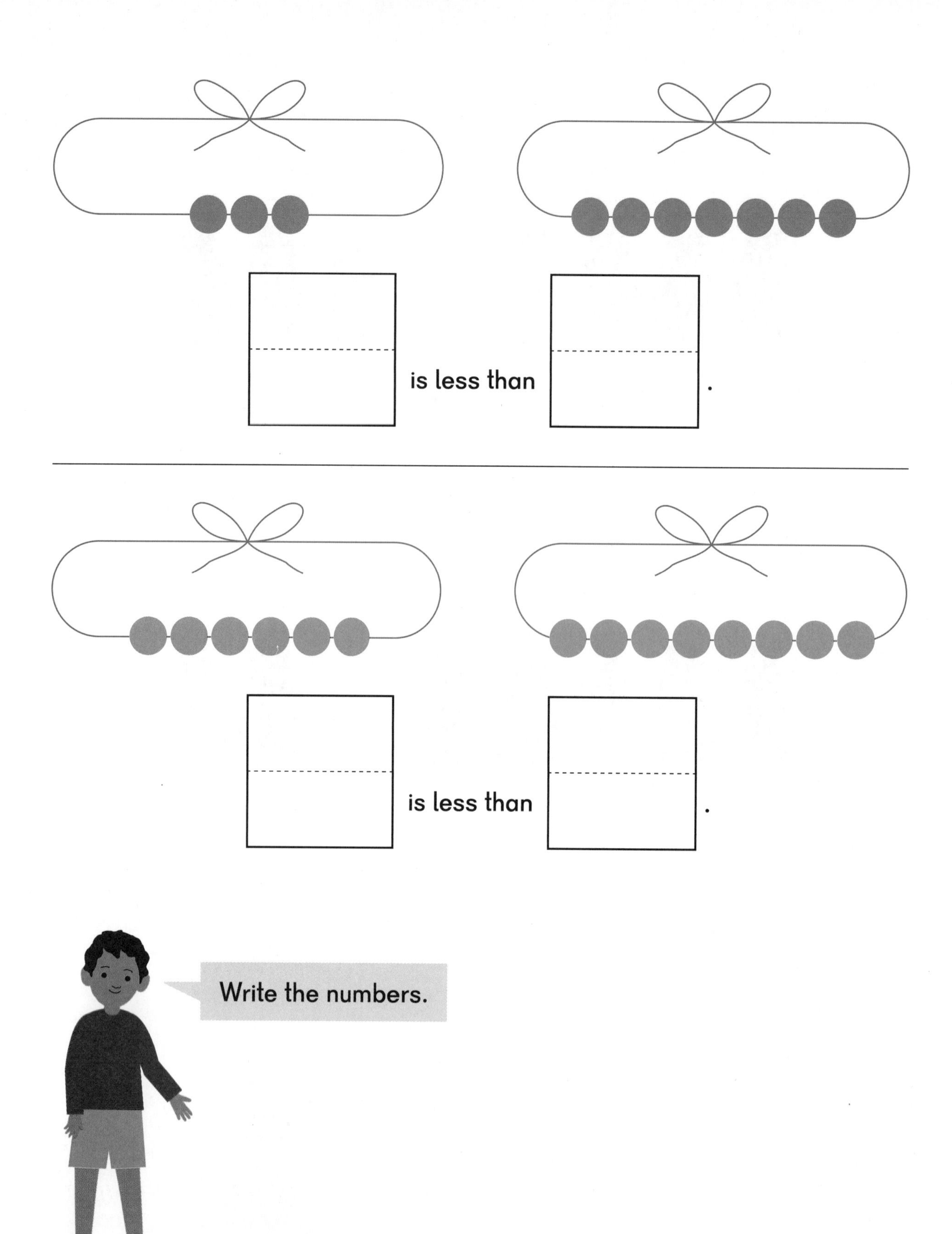

Objective: Compare sets of objects using "less."

Write the numbers.

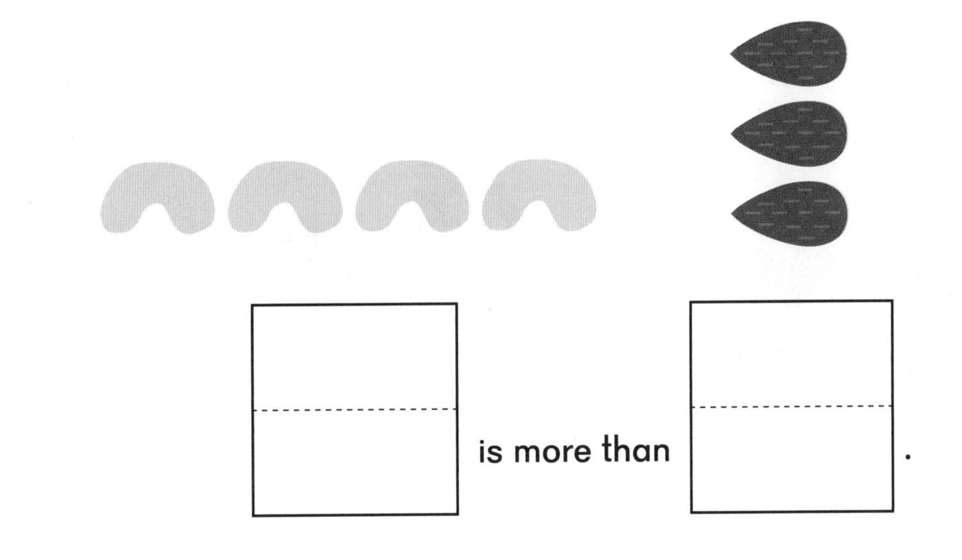

is more than .

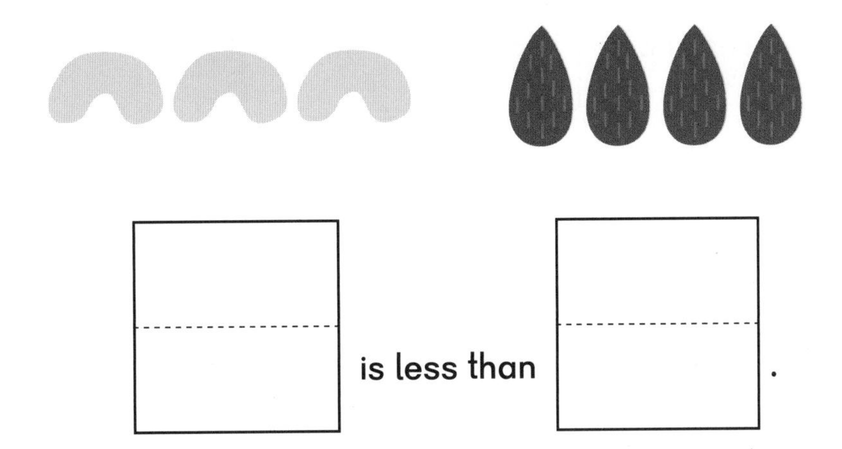

is less than .

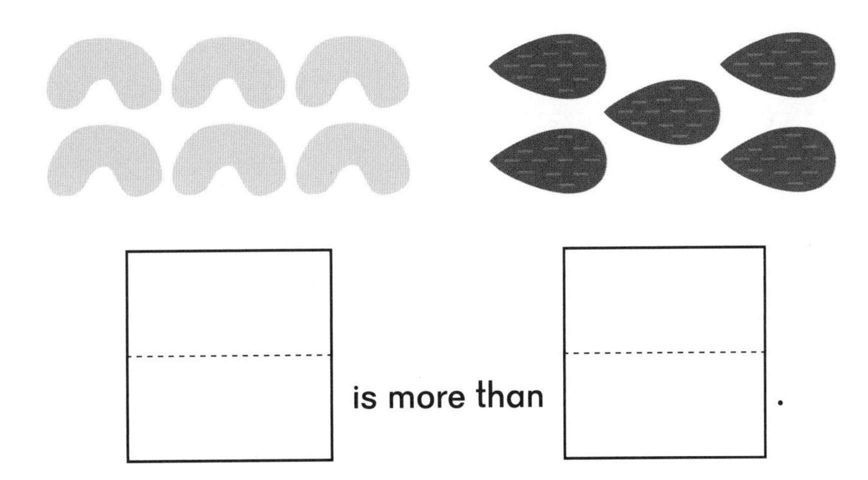

is more than .

Objective: Compare sets of objects using "more" and "less."

Write the numbers.

They need 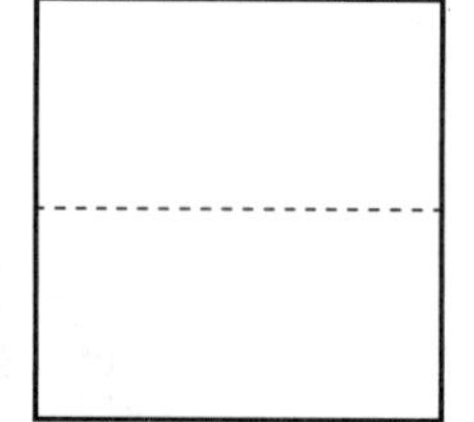 more  .

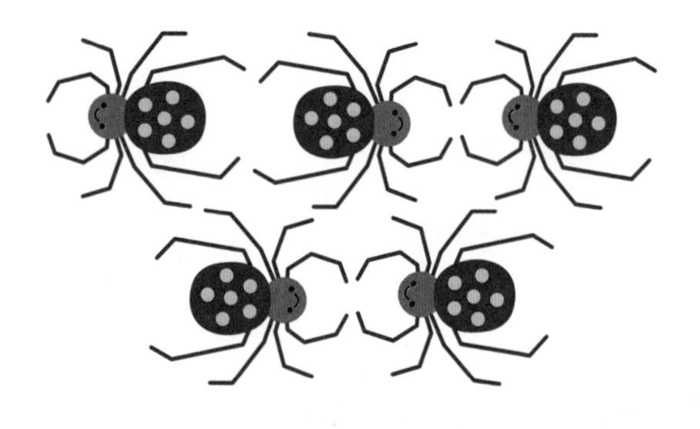

They need 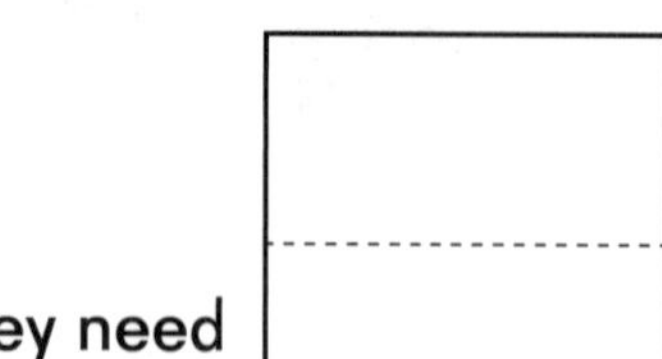 more 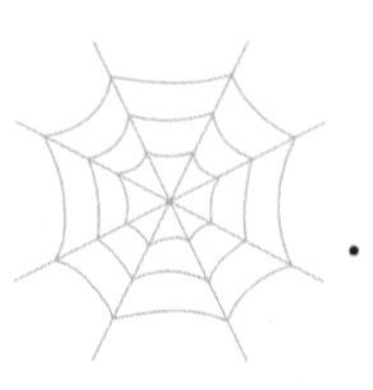 .

Objective: Compare sets of objects and numbers using "more" and "less."

Exercise 3 • page 137

Draw a line to match each straw to a milk shake.

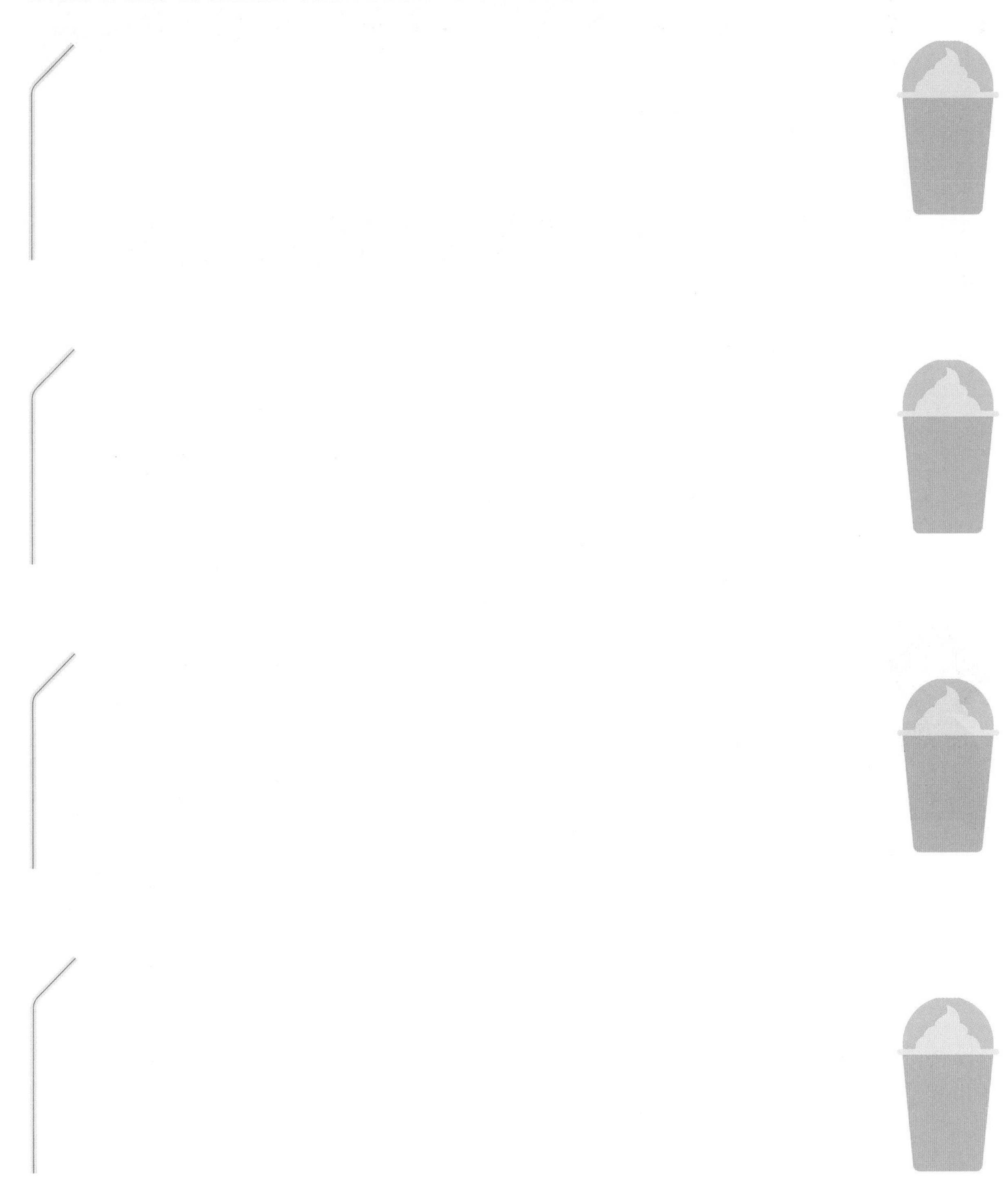

Objective: Practice.

Circle the group that has more.

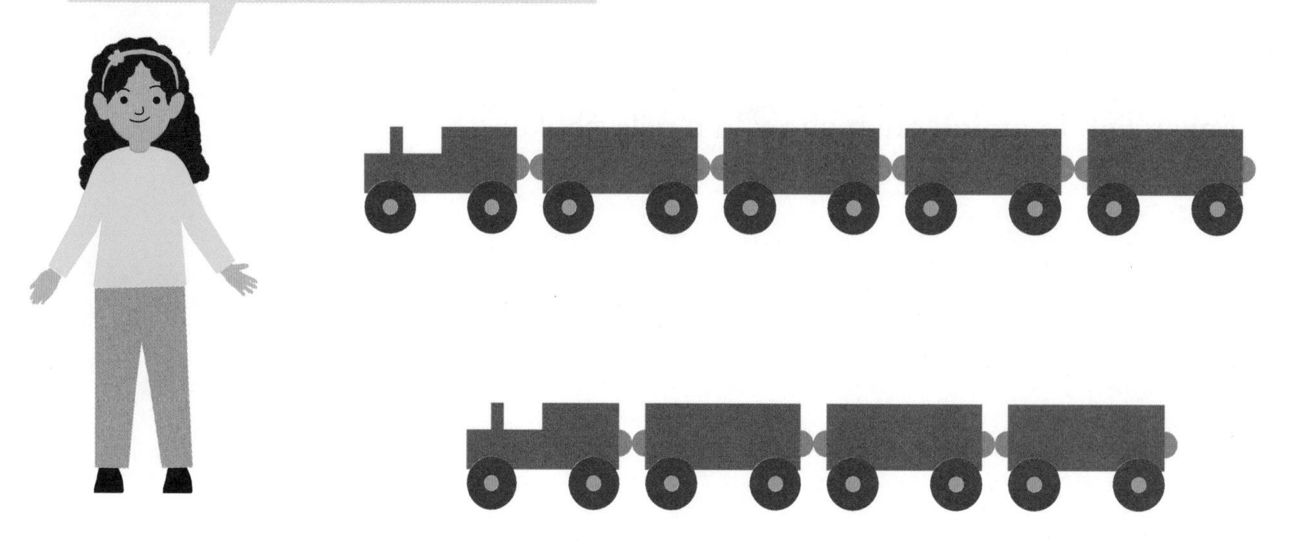

Circle the group that has fewer.

Objective: Practice.

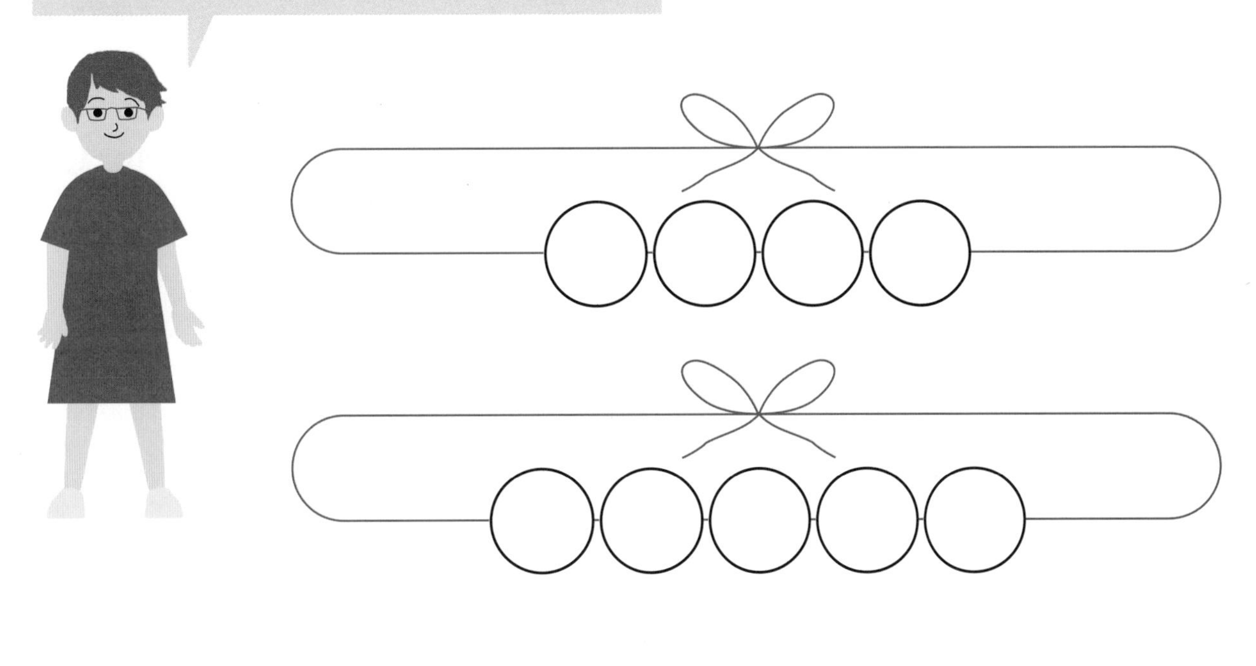

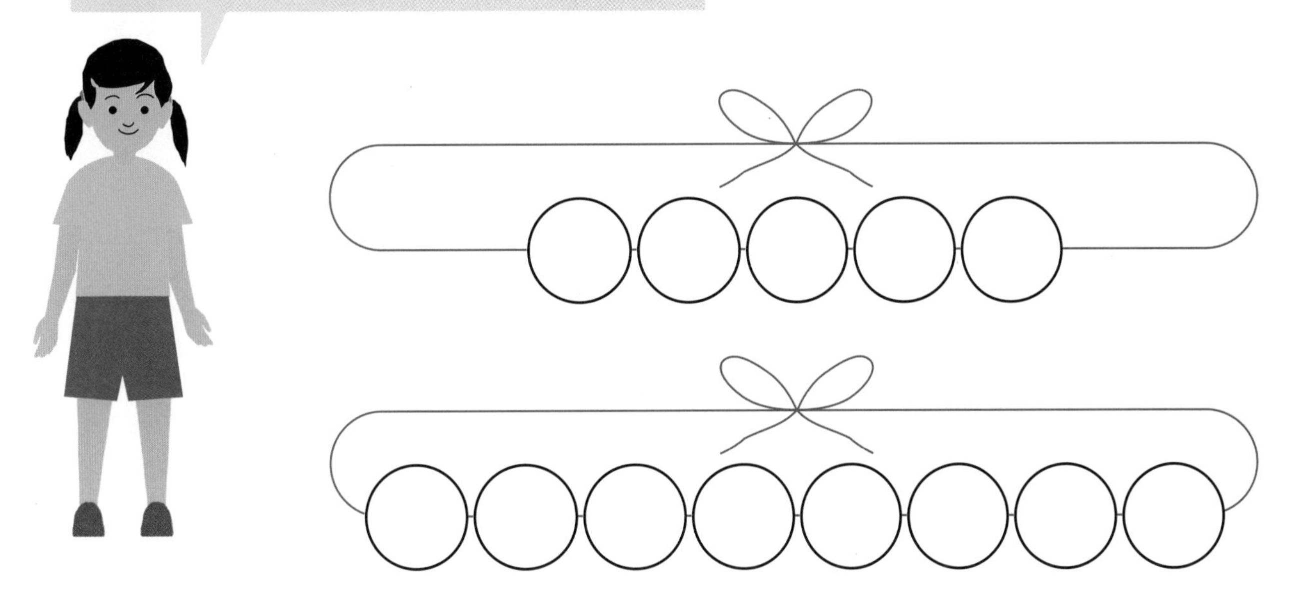

Objective: Practice.

Write the numbers.

Circle the group that has 1 more.

Write the numbers.

Circle the group that has 1 fewer.

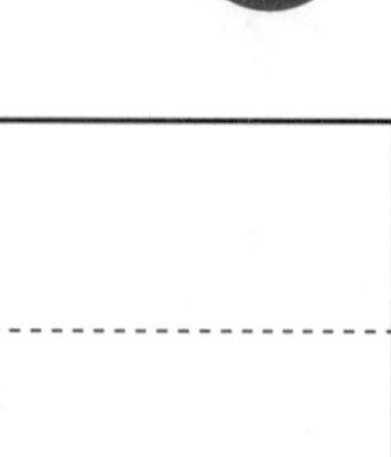

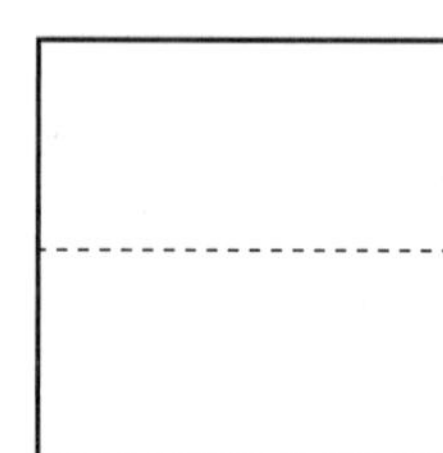

Objective: Practice.

Exercise 4 • page 141

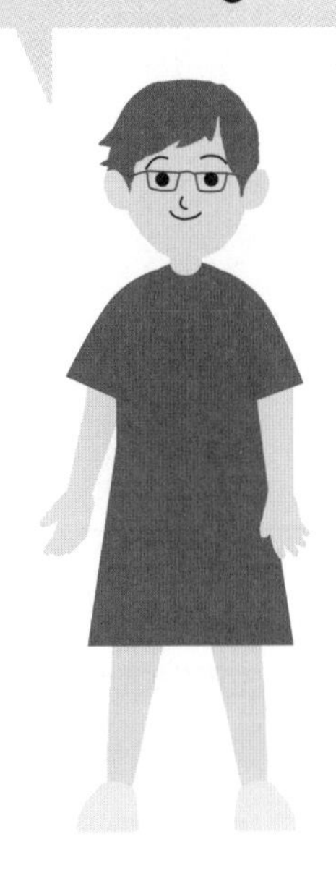

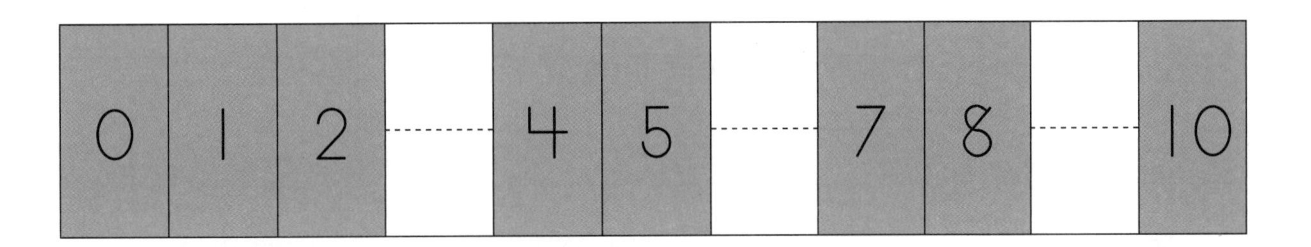

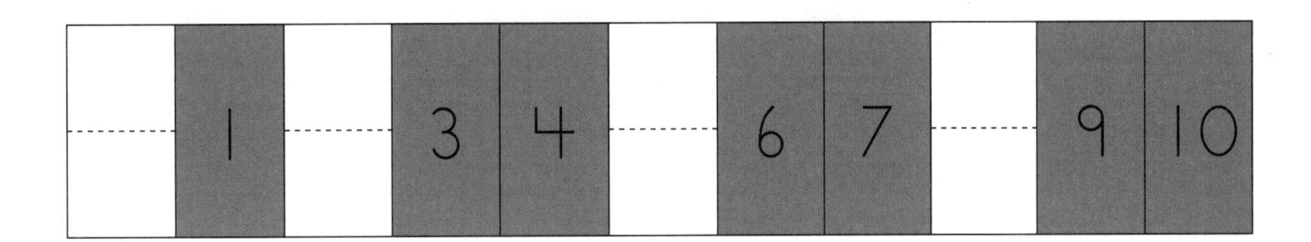

Objective: Practice.

Write the numbers.

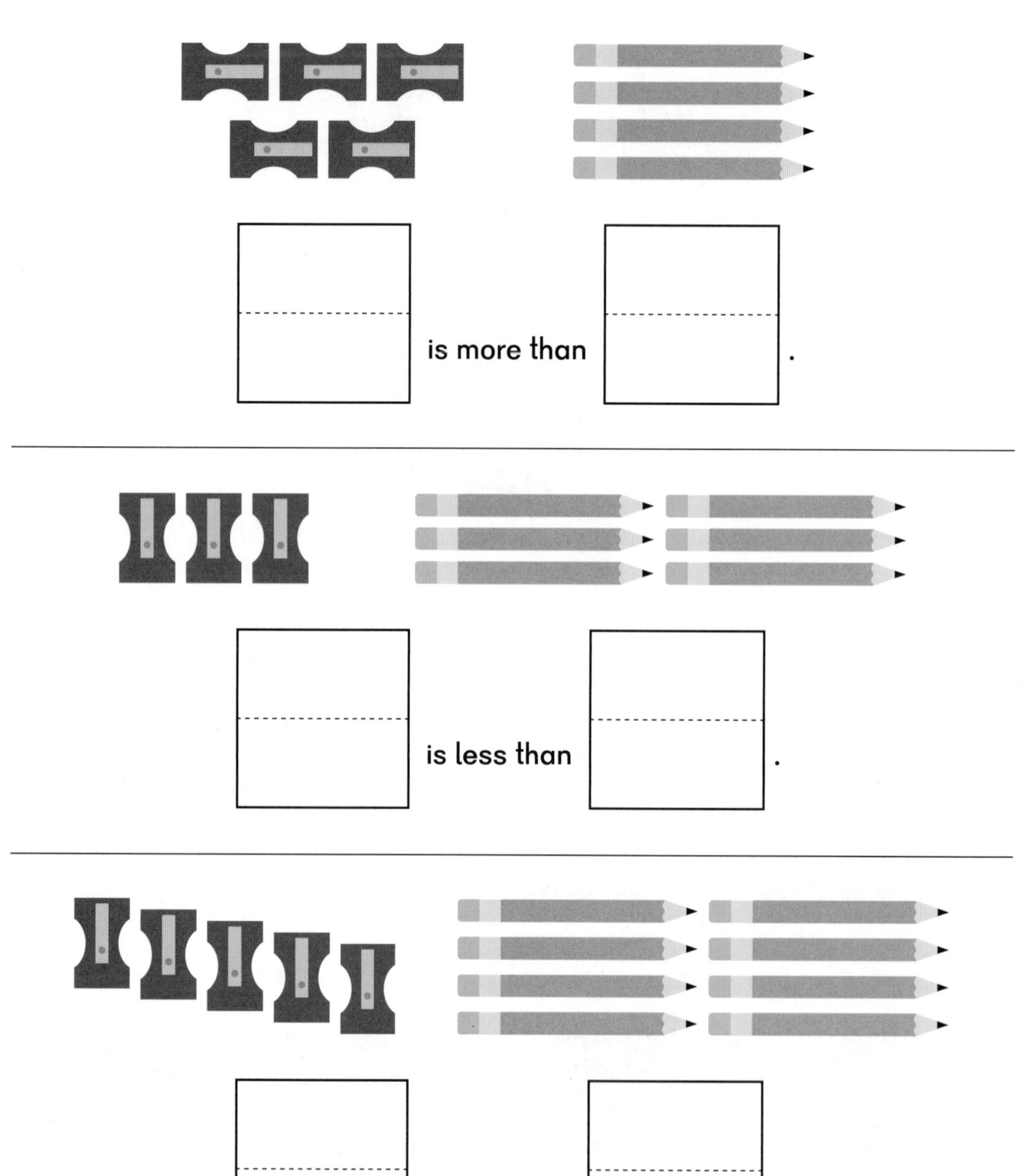

Objective: Practice.

Exercise 5 • page 143

Blank

Blank